DIGITAL SIGNAL PROCESSING FUNDAMENTALS

LIMITED WARRANTY AND DISCLAIMER OF LIABILITY

THE CD-ROM THAT ACCOMPANIES THE BOOK MAY BE USED ON A SINGLE PC ONLY. THE LICENSE DOES NOT PERMIT THE USE ON A NETWORK (OF ANY KIND). YOU FURTHER AGREE THAT THIS LICENSE GRANTS PERMISSION TO USE THE PRODUCTS CONTAINED HEREIN, BUT DOES NOT GIVE YOU RIGHT OF OWNERSHIP TO ANY OF THE CONTENT OR PRODUCT CONTAINED ON THIS CD-ROM. USE OF THIRD-PARTY SOFTWARE CONTAINED ON THIS CD-ROM IS LIMITED TO AND SUBJECT TO LICENSING TERMS FOR THE RESPECTIVE PRODUCTS.

CHARLES RIVER MEDIA, INC. ("CRM") AND/OR ANYONE WHO HAS BEEN INVOLVED IN THE WRITING, CREATION, OR PRODUCTION OF THE ACCOMPANYING CODE ("THE SOFTWARE") OR THE THIRD-PARTY PRODUCTS CONTAINED ON THE CD-ROM OR TEXTUAL MATERIAL IN THE BOOK, CANNOT AND DO NOT WARRANT THE PERFORMANCE OR RESULTS THAT MAY BE OBTAINED BY USING THE SOFTWARE OR CONTENTS OF THE BOOK. THE AUTHOR AND PUBLISHER HAVE USED THEIR BEST EFFORTS TO ENSURE THE ACCURACY AND FUNCTIONALITY OF THE TEXTUAL MATERIAL AND PROGRAMS CONTAINED HEREIN. WE HOWEVER, MAKE NO WARRANTY OF ANY KIND, EXPRESS OR IMPLIED, REGARDING THE PERFORMANCE OF THESE PROGRAMS OR CONTENTS. THE SOFTWARE IS SOLD "AS IS" WITHOUT WARRANTY (EXCEPT FOR DEFECTIVE MATERIALS USED IN MANUFACTURING THE DISK OR DUE TO FAULTY WORKMANSHIP).

THE AUTHOR, THE PUBLISHER, DEVELOPERS OF THIRD-PARTY SOFTWARE, AND ANYONE INVOLVED IN THE PRODUCTION AND MANUFACTURING OF THIS WORK SHALL NOT BE LIABLE FOR DAMAGES OF ANY KIND ARISING OUT OF THE USE OF (OR THE INABILITY TO USE) THE PROGRAMS, SOURCE CODE, OR TEXTUAL MATERIAL CONTAINED IN THIS PUBLICATION. THIS INCLUDES, BUT IS NOT LIMITED TO, LOSS OF REVENUE OR PROFIT, OR OTHER INCIDENTAL OR CONSEQUENTIAL DAMAGES ARISING OUT OF THE USE OF THE PRODUCT.

THE SOLE REMEDY IN THE EVENT OF A CLAIM OF ANY KIND IS EXPRESSLY LIMITED TO REPLACEMENT OF THE BOOK AND/OR CD-ROM, AND ONLY AT THE DISCRETION OF CRM.

THE USE OF "IMPLIED WARRANTY" AND CERTAIN "EXCLUSIONS" VARIES FROM STATE TO STATE, AND MAY NOT APPLY TO THE PURCHASER OF THIS PRODUCT.

DIGITAL SIGNAL PROCESSING FUNDAMENTALS

ASHFAQ A. KHAN

DA VINCI ENGINEERING PRESS
Hingham, Massachusetts

Copyright 2005 by DA VINCI ENGINEERING PRESS, an imprint of CHARLES RIVER MEDIA, INC.
All rights reserved.

No part of this publication may be reproduced in any way, stored in a retrieval system of any type, or transmitted by any means or media, electronic or mechanical, including, but not limited to, photocopy, recording, or scanning, without *prior permission in writing* from the publisher.

Editor: David Pallai
Cover Design: The Printed Image

DA VINCI ENGINEERING PRESS
CHARLES RIVER MEDIA, INC.
10 Downer Avenue
Hingham, Massachusetts 02043
781-740-0400
781-740-8816 (FAX)
info@charlesriver.com
www.charlesriver.com

This book is printed on acid-free paper.

Ashfaq A. Khan. *Digital Signal Processing Fundamentals.*
ISBN: 1-58450-281-9

All brand names and product names mentioned in this book are trademarks or service marks of their respective companies. Any omission or misuse (of any kind) of service marks or trademarks should not be regarded as intent to infringe on the property of others. The publisher recognizes and respects all marks used by companies, manufacturers, and developers as a means to distinguish their products.

Library of Congress Cataloging-in-Publication Data
Khan, Ashfaq A.
  Digital signal processing fundamentals / Ashfaq A. Khan.— 1st ed.
    p. cm.
  Includes index.
  ISBN 1-58450-281-9 (hardback with cd-rom : alk. paper)
  1. Signal processing—Digital techniques. I. Title.

TK5102.9.K3365 2005
621.382'2—dc22
                                    2004020495

Printed in the United States of America
04 7 6 5 4 3 2 First Edition

CHARLES RIVER MEDIA titles are available for site license or bulk purchase by institutions, user groups, corporations, etc. For additional information, please contact the Special Sales Department at 781-740-0400.

Requests for replacement of a defective CD-ROM must be accompanied by the original disc, your mailing address, telephone number, date of purchase, and purchase price. Please state the nature of the problem, and send the information to CHARLES RIVER MEDIA, INC., 10 Downer Avenue, Hingham, Massachusetts 02043. CRM's sole obligation to the purchaser is to replace the disc, based on defective materials or faulty workmanship, but not on the operation or functionality of the product.

I dedicate this book to the LIGO project.

Contents

	Acknowledgments	xvii
	Preface	xix
1	**Fourier Analysis**	**1**
	Continuous Time and Discrete Time	2
	Periodic Waves	3
	A Simple Periodic Function	3
	Mathematical Description of a Wave Function	4
	Fourier Synthesis	11
	Fourier Analysis	16
	Example 1.1	16
	Fundamental Frequency	18
	Harmonics	19
	Nyquist Frequency	20
	Orthogonal Functions	20
	Example 1.2	27
	Summary	33
2	**Complex Number Arithmetic**	**35**
	Complex Number Representation	36
	The Imaginary Operator $\sqrt{-1}$	36
	Complex Conjugate	38
	Complex Numbers in Polar Coordinates	40
	Powers of Complex Numbers	42
	Roots of Complex Numbers	43

The Exponent *e* and the Power Functions	43
Euler's Identity	47
The Complex Frequencies	48
Conjugate Function	50
The Phasor Method	51
The Electrical Networks	51
Example 2.1	54
Example 2.2	54
Example 2.3	55
Example 2.4	56
Example 2.5	57
Example 2.6	58
Example 2.7	58
Example 2.8	59
Example 2.9	59
Example 2.10	60
Example 2.11	60
Example 2.12	61
Resonance	62
Example 2.13	63
Example 2.14	64
Signal Modulation	64
Heterodyne	66
Summary	66
3 The Fourier Transform	**69**
A Periodic Function as a Complex Number Function	70
Conventions and Notations	74
The Fourier Series of a Periodic Pulse Train	75
Removing the Periodic Dependency	76
Euler's Equation and Nonperiodic Waves	77

Fourier Transforms 77
   The Difference Between Fourier Transforms and Fourier Series 81
   The Difference Between a True Function and the Estimation with the Fourier Transform 81
   Discrete Time Fourier Transforms 85
Fast Fourier Transforms 86
   Vector Rotation 91
   FFT Algorithm 92
Random Signals 93
   Amplitude Distribution 93
   Autocorrelation 94
   Cross Correlation 95
   Power Spectrum 96
   White Noise, Pink Noise, and 1/f Noise 97
Summary 99

4 Solutions of Differential Equations 101

Linear Time Invariant Systems 102
   Differential Equations 103
   Systems Response 104
   Solutions of Differential Equations 104
   Linear Differential Equations 104
   The General Form of a Differential Equation 105
First Order Differential Equations 106
   Natural Response 106
   Forced Response 108
   Step Response 109
   Unit Impulse Response 112
   Scaled Impulse Response 116
The Convolution Process 117
   Discrete Time Convolution 117

	Second Order Differential Equations	123
	General Form of the Second Order Differential Equations	126
	Natural Response	127
	Roots Real and Distinct	131
	Roots Real and Equal	133
	Roots Complex	134
	Forced Excitations	137
	Step Response	137
	The Complex Plane	141
	Unit Impulse Response	143
	Scaled Impulse Response	146
	Response to an Arbitrary Input	146
	Summary	146
5	**Laplace Transforms and z-Transforms**	**147**
	The Laplace Transform	148
	The Linearity Property of the Laplace Transform	149
	Some Useful Transformations	150
	The Rules of Differentiation	151
	The Rules of Integration	152
	The Inverse Laplace Transform	153
	Solving Differential Equations with the Laplace Transform	153
	Impedance and Admittance	157
	Transfer Functions	160
	Filter Design and the Transfer Function	162
	Reducing a Transfer Function's Complexity	166
	Poles, Zeros, and Steady State Frequency Response	166
	Decibel dB	170
	The Bode Plot	171
	The z-Transform	173
	Inverse z-Transforms	174

Unit Impulse Functions   175
Unit Step Functions   175
Exponential Functions   175
Sine Functions   175
Cosine Functions   176
The Difference Rule   176
The Transfer Function in z-Transforms   177
Poles and Zeros of z-Transforms   180
Summary   182

6 Filter Design   183

Filter Terminologies   184
Filter Design Methods   186
    Frequency-frequency Mapping   187
    Building Blocks of Transfer Functions   189
    Butterworth Filters   197
Analog Filters   201
    Active *RC* Filters   205
Example 6.1   208
Example 6.2   208
Example 6.3   209
Example 6.4   210
Example 6.5   213
Example 6.6   215
Example 6.7   215
Example 6.8   217
Summary   219

7 Digital Filters   221

Digital Filter Design Process   222
    The Convolution Sum   223

The Frequency Variable $\omega \Delta t$	224
Digital Filter Transfer Functions	226
The Transfer Function Frequency and Phase Response	228
Transfer Function Models	228
Rational Polynomial Transfer Functions	228
Transfer Functions with Poles-Zero-Gain	230
Gain	231
Digital Filter Characterization	232
Low-Pass Filters	232
Example 7.1	233
High-pass Filter	234
Example 7.2	235
The Band-Pass and the Band-Stop Filters	236
Band-Pass Filters	237
Example 7.3	238
Band-Stop Filters	239
Example 7.4	240
Example 7.5	241
Realizing the Filter Response	243
The Coefficients of the Difference Equation	243
Cascading Transfer Functions	245
Low-Pass Filter Realization	246
High-Pass Filter Realization	247
Example 7.6	247
Improving Filter Response	249
Butterworth Filters	250
Example 7.7	251
Example 7.8	251
Example 7.9	252
Example 7.10	253
Chebyshev Filters	254

Example 7.11	257
Example 7.12	258
Example 7.13	258
Example 7.14	259
Chebyshev Filters with Ripples in Stop Band	260
Example 7.15	260
Elliptic Filters	261
Example 7.16	262
The Transition from Analog to Digital Filters	263
Example 7.17	264
Example 7.18	265
Example 7.19	265
Summary	267

8   The FIR Filters   269

Ideal Filters	270
The Fourier Series Representation of the Ideal Frequency Response	271
FIR Filter Design	274
Low-Pass FIR Filter	275
Example 8.1	276
High-Pass Filters	280
Example 8.2	281
Band-Pass Filters	282
Example 8.3	283
Band-Stop Filters	284
Example 8.4	285
Windowing	287
The Rectangular Window	290
Example 8.5	292
The Power Window	293

Example 8.6		297
The Von Hann Window		297
The Hamming Window		300
Summary		306

Appendix A   Matlab Tutorial   307

Executing Scripts	308
Importing Data from a File	308
Exporting Data to a File	308
Operations on Matrices	309
Complex Numbers	313
The Colon Operator	315
Expression and Special Constants	316
Other Data Structures	316
Multidimensional Arrays	317
Cell Arrays	317
Characters and Text	317
Conversion	317
Structures	318
Functions	318
`eval`	318
Polynomials	318
Evaluating Polynomials	319
Systems of Linear Equations	320
The \ and / Division Operators	321
Polynomial Least Square Fit	322
Polynomial Regression	323
Flow Control	324
Signal Processing	326
Generating Waveforms	326
Convolution	328

Transfer Functions	328
Zero-Pole-Gain (zpk)	328
State-Space	328
Partial Fraction Expansion (Residue Form)	329
Second Order Section (SOS)	329
Lattice Structures	329
Discrete Fourier Transform	329
Filter Design	330
Removing Phase Distortion from a Filter	330
Impulse Response	330
Frequency Response Bode Plot	330
Delay	331
Pole-Zero Analysis	331
IIR Filters	331
Chebyshev Type 1 Filters	331
Chebyshev Type II Filters	332
Butterworth Filters	332
Elliptic Filters	332
The Windowing Method and FIR Filters	333
The Hamming Window	333
The Von Hann Window	333
Appendix B Scilab Tutorial	**335**
Scilab Basics	335
Polynomial Operations	336
Evaluation of Polynomials	338
Plotting	339
Example B.1	340
Example B.2	340
The Transfer Function	341
Convolution	342

Example B.3	342
Fast Fourier Transforms	343
Correlation	343
Example B.4	343
IIR Filters	344
Example B.5	345
Transfer Function Magnitude	346
Poles and Gain	347
Butterworth Filters	349
Window FIR Filters	350

Appendix C   Digital Filter Applications   351

Image Processing	351
Point Operations	352
Neighborhood Operations	354
Morphological Operations	357
Edge Detection	357
Audio Processing	358
DTMF Signal Detection	359
filterTest.pl: Testing the Filter Design	361

Appendix D   About the CD-ROM   365

Appendix E   Software Licenses   371

Appendix F   Bibliography   381

Index   383

Acknowledgments

First, I would like to thank the manuscript reviewers, especially Dr. Frank Candocia of Florida International University for his valuable input. His in-depth critical review of the manuscript helped me to logically organize the contents of the book. I would also like to thank the scientists at LIGO Livingston Observatory for their kind help and explanations of the application of DSP techniques in the Laser Interferometer Gravitational Wave detection apparatus.

Finally, I would like to thank my editor, David Pallai, for his support and guidance. This book would not have been possible without his encouragement.

Preface

DSP, or Digital Signal Processing, is the analysis and processing of physical phenomena that can be measured and digitized using a digital computer. The phrase "physical phenomena" is being used in a broad manner here: it may be a voice over the telephone, an image from a camera, a retinal scan for identification, a thumbprint search, a tracking radar, the monitoring of a heart rate, the servo mechanism of a machine, weather prediction, listening to the tiny ripples from far-off galaxies, or thousands of other instances in which events are sampled, digitized, and passed through a computer for specific processing. It used to fall to an analog system to perform most of the processing, as computers were either very expensive or too slow to handle the demanding calculations required by DSP algorithms.

It is the recent advancement in computer design and speed that has given us the opportunity to apply the techniques of DSP to a vast array of applications which would have been unthinkable just a few years ago. DSP, with its inherent flexibility, is fast replacing the analog designs that were rigid and difficult to operate. The crispness and clarity of a voice over the phone carrying from thousands of miles away is only possible because we use DSP every step of the way, starting from compressing the voice, then transmitting it via satellite, filtering noise, amplifying the sound to an audible level, and finally canceling the echo. Image processing tasks including rotation, scaling, translation, enhancement, sharpening, and morphological warping of an image can all be performed by the home computer thanks to the efficiency of DSP.

The filtering of unwanted frequencies from an input waveform is probably the most common application of DSP; thus it replaces hardware components such as inductors, resistors, and capacitors of electrical circuits that used to perform such operations. In its simplest form, we use DSP to process a set of data such as Dow

Jones Industrials to find the most accurate average curve over a period of time. All this processing requires looking at data through a convolving window where the input is translated by the characteristics of the window and we only see the output produced by the convolving operation.

The purpose of this book is to explain the fundamentals of the DSP techniques that enable you to apply DSP to an existing application or to find a new application that can benefit from such an excellent toolset. The book also provides material of academic interest, because a mathematical foundation is presented to explain the concepts behind the DSP techniques. You will also gain an appreciation for the mathematical geniuses of Fourier, Laplace, and Euler, who created the foundation for modern-day DSP two centuries ago with only their imaginations.

Although the algorithms of DSP are very simple and straightforward, they require not only a strong understanding of their mathematical foundations, but a change of mindset. One must learn to think of a set of data as a function made up of component frequencies. It is hard to imagine that a straight line (in a domain of interest) is actually composed of superimposed simple periodic waves, but that is exactly what Fourier proved. He then devised a method of finding the component frequencies in a given set of data so that a mathematical operation could be performed upon them without requiring a prior knowledge of the type of function that represents the data.

The book is organized into eight chapters. Chapter 1 formulates the Fourier series that forms the basis of DSP. Chapter 2 is devoted to the mathematics of complex numbers and vector algebra, providing a mathematical foundation for operations on periodic functions. Fourier Transforms are introduced in Chapter 3, which also explains the algorithms for Discrete Fourier Transforms and Fast Fourier Transforms. Chapter 4 introduces the solution of differential equations through convolution. Chapter 5 explains the process of Laplace Transforms and z-Transforms. Chapter 6 discusses the fundamentals of filter design, covering the basics of analog filtering techniques. Chapter 7 covers digital filtering with an emphasis on IIR filters and Chapter 8 covers FIR filters and windowing of input data, including Power, Hamming, Hanning, and Kaiser Windows. Appendixes A and B are short tutorials on the *Matlab* and *Scilab* software applications. The graphs and tables presented in the book were designed using scripts; the listings are presented on the accompanying CD-ROM. Appendix C provides three short application programs covering digital image processing, digital audio processing, and IIR filtering. The applications are based on Perl scripts using a public-domain graphing utility. This source code is also provided in the accompanying CD-ROM.

The intended audience for this book includes practicing electrical engineers, software engineers, and engineering students who have an interest in DSP techniques. No specific prior knowledge is assumed other than some basic familiarity with calculus, which is required to understand the derivatives used in the text.

Ashfaq A. Khan
Baton Rouge, LA

1 Fourier Analysis

In This Chapter

- Continuous Time and Discrete Time
- Periodic Waves
- Fourier Synthesis
- Fourier Analysis

We begin our study of digital signal processing (DSP) with an introduction to Fourier analysis. Although a somewhat complex place to start, it is necessary to the discussion of physical signals that are measurable, such as temperature, pressure, voice, and sound. These signals, though arbitrary in nature, must be defined as mathematical functions if further processing is required; for example, removing the unwanted noise from the music or designing a control system for maintaining a constant action. The central idea of the Fourier analysis is to look through these events as if they are functions made up of superimposed sinusoidal frequencies. How to identify these component frequencies is the topic of discussion in this chapter.

The seventeenth century mathematician Jean Baptiste Joseph Fourier (1768–1830) discovered that a continuous time periodic function (one that repeats a pattern periodically after a certain interval) could be approximated as a series of simple sin and cosine functions. In his honor, such an approximation of the function is called the *Fourier series* of the function. Although it is hard to find functions in nature that are truly periodic for all time, we could still define a proper domain of interest and approximate a function using its Fourier series. When we deal with an arbitrary signal and we don't know its period (see Chapter 3, "The Fourier Transform"), we can remove this periodic dependency: we obtain what we call the relative magnitudes of the component frequencies, and that becomes the Fourier Transform of the signal. In essence, Fourier analysis provides us the domain of frequency to analyze a signal.

Frequency analysis is important in many situations, as systems behave differently at different frequencies. Imagine a child on a swing; the swing is a system and it requires an external periodic force for its operation. Only when the frequency of the external force matches the natural frequency of the swing will the swing keep going, otherwise, it will stop. If we knew the Fourier series of the external force beforehand, we could predict whether the swing will move and for how long. When the component frequencies of a function are identified, the overall processing of the system could be evaluated as the sum of the responses of the individual frequencies, as if each were a separate input applied to the system. For some classes of functions, further processing is only possible once the component frequencies are identified; this is evident from the example presented at the end of the chapter.

The Fourier series and Fourier Transform are not the only aspects of DSP, and it is not necessary to transform all functions into frequency components, but the concept is important and you should be able to deal with it when it is necessary. In this chapter, we will develop techniques for analyzing and formulating a waveform as described by Fourier.

CONTINUOUS TIME AND DISCRETE TIME

In every field of engineering, we take signals and process them to produce a desired outcome. In most cases, there is an analog signal to be acquired from a transducer and processed (the transducer may have converted some other form of energy to electrical form). The processing involves integrating, differentiating, smoothing, and filtering the signals as well as acquiring data for display and archive purposes. The processing may be performed using analog circuitry, or we may use a computer to perform the required mathematical calculations and produce the outcome.

Analog signal processing is defined by continuous time as the signals flow continuously through the analog circuit components, and the output is generated without a time delay. In the context of digital signal processing, when we use a computer to process a signal, there is always a finite amount of time between acquisition and subsequent processing, no matter how fast the computer. The method is essentially a sampling of an event taken at a discrete time. The mathematics has to be modified slightly to take into account that only at specific times does the discrete time coincide with the real time.

If Δt is our sampling period, the *kth* sample time that coincides with the real time is $t = k\Delta t$. We usually represent a function of continuous time as $f(t)$, and a function sampled in discrete time is $f(k)$ or $f(n)$.

Throughout the book, we will be using the terminology continuous time and discrete time concurrently and showing the difference in the two mathematical formulations accordingly, as we develop the applications.

PERIODIC WAVES

By definition, a function is periodic if its basic pattern is repeated with a predefined time interval or period. A period is simply the time between each crust of the wave and is fixed for a true periodic wave. Fourier showed us that a seemingly arbitrary waveform is basically composed of several simple sinusoidal waves. Further, he not only showed us how to compute the exact number of these simple waves but also how to extract the height of each individual wave.

(Note that the word *wave* is being used in the context of sinusoidal waves and not the traveling waves at a distance.)

A Simple Periodic Function

Physical events may be classified as periodic or nonperiodic. Any repeated pattern could be considered a periodic function, but a simple periodic function is a cyclic pattern that has fixed amplitude and a fixed period, as shown in Figure 1.1. Think of a bicycle with a painted spot on its wheel. The graph of the spot as the wheel moves may be described as shown in Figure 1.3. The path traversed by the spot on the wheel is an example of a simple wave. When the wheel makes a complete revolution, the spot achieves a peak and a valley and comes back to its original height. As time progresses, the same pattern is repeated. It starts at 0°, achieves a full height at 90°, goes back to 0 height at 180°, at 270° it reaches its minimum height, and then it goes back to 0 at 360°.

The two important aspects are the maximum height and the period of revolution. For a simple wave, the amplitude and the period are fixed. The period tells us the time it takes to make a complete revolution, and the amplitude tells us the maximum height the wave can achieve. The inverse of period is frequency indicating how many revolutions are completed in one second. This is a typical cyclic function: *x* direction is the angle of rotation, or the independent variable, and *y* direction is the amplitude, or the dependent variable. Algebraically speaking, a periodic function is a function of angular displacement in time and the amplitude of

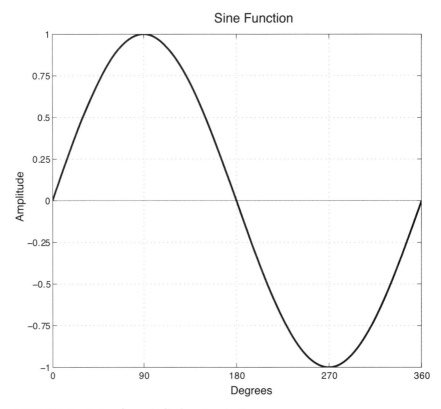

FIGURE 1.1 A simple periodic function in time.

the wave. The amplitude attains one maximum high and one maximum low during one complete cycle.

Mathematical Description of a Wave Function

Before we begin to describe an arbitrary wave, let's discuss how a simple sinusoidal wave is being formed as a function of time. A function is a relationship of two or more quantities described algebraically in a mathematical form. In our previous example of the motion of the spot on the bicycle wheel, there is a direct relationship between the height of the spot and the rotation of the wheel. Angles are measured in degrees, if a circle is sliced into 360 equal parts, each is 1° of rotation, while angles are measured in radians if the arc of rotation is equal to the length of its radius itself. There are exactly 2π radians in one complete cycle, as shown in the Figure 1.2. One radian is slightly less than 60°.

Fourier Analysis

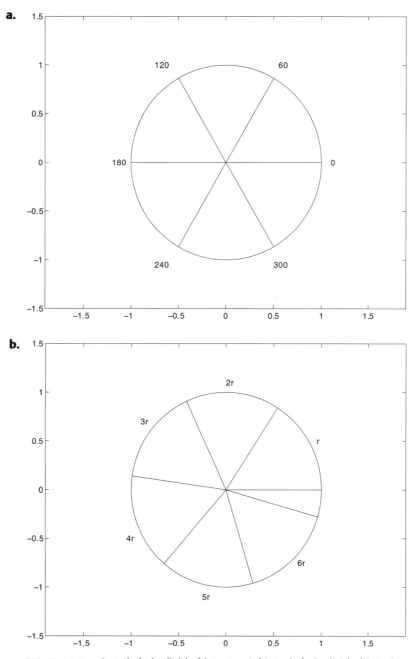

FIGURE 1.2 a) A circle is divided into 360°. b) A circle is divided into 2π radians.

Going back to the bicycle example, the motion of the spot can be described in two ways:

- The change in the height of the spot with respect to the angular movement of the wheel.
- The change in the position of the spot with respect to time.

There are three variables in our function: the height h, the angular distance θ, and the time t. First, let's develop the functional relationship between the height and the angular displacement θ.

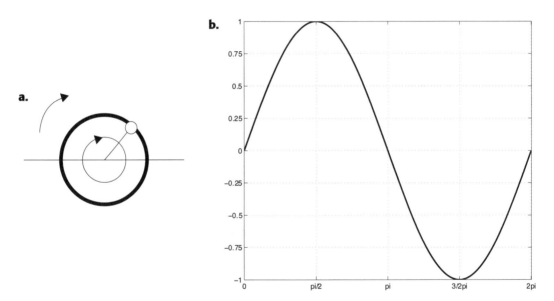

FIGURE 1.3 a) Wheel with a spot. b) Motion of the spot as the wheel moves in time.

Sin and Cos Functions

Looking at the rotation of the bicycle wheel starting from 0 angle, the spot is 0 feet high; the change in the pattern of height h with respect to angle θ is shown in Figure 1.4.

The spot gradually increases in height as we increase the angle θ, until it reaches its full height of 1 foot that is the radius of the wheel, exactly at $\pi/2$ radians, further increase in angle brings down the height until it reaches 0 at π radians, completing the half cycle. Further increase in the angle results in an increase in height but in the

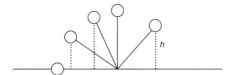

FIGURE 1.4 Height *h* as a function of the angle θ.

opposite direction, reaching maximum negative height at $3\pi/2$ radians. From $3\pi/2$ radians to 2π radians, the height comes back to 0, where we started. The cycle is repeated forever as the angle is increased further. Trigonometrically, the relationship of the height *h* and the angular displacement θ can be described as a sine function, as shown in Figure 1.5.

$$h = f(\theta)$$

$$h = r \times \sin(\theta)$$

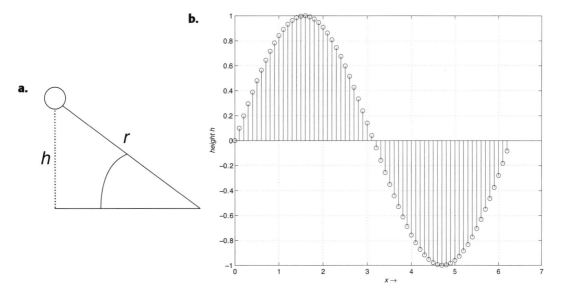

FIGURE 1.5 a) Height of the spot as a sine function of the angle θ. b) Projection on the x axis.

If we start the cycle when the height is at the top, the trigonometric relationship is defined in terms of a cosine function, as shown in the Figure 1.6.

$$h = r \times \cos(\theta)$$

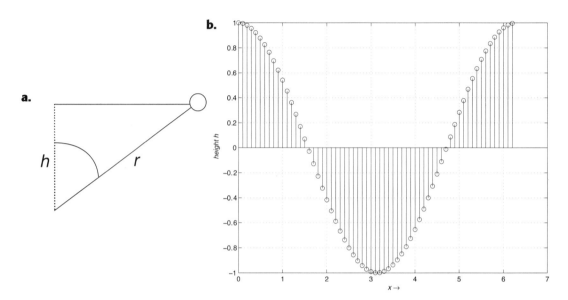

FIGURE 1.6 a) Graph of the cosine function, showing height versus angle. b) Height h as a cosine function of the angle θ.

Sine and cosine functions are identical except the pattern of the cosine function is delayed by 90°. The angular difference between two functions is called *phase delay*; the phase delay between sine and cosine in this case is 90°.

Angular Velocity and the Frequency of a Periodic Function

Looking at the motion of the bicycle wheel, the angular displacement in itself is a function in time. It could be described as frequency f with units of cycles-per-second or angular velocity ω as radians/sec. The *amplitude* of a wave is its maximum height, which is essentially the radius r of the circle formed by the circular motion.

Frequency

Frequency is defined as:

$$f = cycle/sec$$

If the period of one complete cycle is T seconds, then

$$f = \frac{1}{T}$$

Angular Velocity

When velocity is measured as change of angle per unit time,

$$\omega = \frac{\theta}{t}, radians/\sec$$

$$\theta = \omega t, radians$$

A complete cycle is 2π radians, thus the angular velocity in radians is given as,

$$\omega = 2\pi f$$

$$\omega = \frac{2\pi}{T}$$

Substituting the angle θ for the angular displacement in time $\theta = \omega t, radians$, we can describe the mathematical formulation of a simple periodic wave in terms of a sine function as well as a cosine function,

$$h = r \times \sin(\omega t)$$

$$h = r \times \cos(\omega t)$$

If the angular displacement is described in frequency, we must convert it into radians for further calculations,

$$h = r \times \sin(2\pi f t)$$

$$h = r \times \cos(2\pi f t)$$

If we assume the bicycle is moving with a velocity of 1 ft/sec and the spokes of the wheel are 1 foot long, a complete revolution of 2π feet is achieved in 2π seconds, or the angular velocity ω of 1 radian per second or the frequency $\frac{1}{2\pi}$ cycle /sec.

$$r = 1$$

Digital Signal Processing Fundamentals

$$\omega = \frac{\theta}{t} = 1\,radian/\sec$$

$$f = \frac{1}{2\pi}\omega = 0.159\,cycles/\sec$$

When we use computers to process data, it is no longer continuous time processing. The time t is valid only at the instance when we have acquired the sample data. If k is the sample number and Δt is the time interval per sample, the discrete time variable equivalent of continuous time t is $k\Delta t$. Thus, the continuous time frequency function $\sin(\omega t)$ equals the discrete time frequency function $\sin(\omega k\Delta t)$; this can also be defined in terms of the period of the waveform by substituting $\omega = \frac{2\pi}{T}$.

Discrete Time Cyclic Frequencies

Discrete time cyclic frequencies are defined as follows:

$$\sin(2\pi k\Delta t/T)$$

$$\cos(2\pi k\Delta t/T)$$

Properties of Simple Periodic Waves

The two distinctive properties of simple periodic functions are the amplitude h and the angular velocity ω. A specific simple wave is different from all other simple waves if its amplitude and the angular velocity ω are different from all others. It is customary to differentiate waves by using subscript variable r and ω.

$$f_1(t) = r_1 \times \sin(\omega_1 t)$$
$$f_2(t) = r_2 \times \sin(\omega_2 t)$$
$$f_3(t) = r_3 \times \sin(\omega_3 t)$$
$$f_4(t) = r_4 \times \sin(\omega_5 t)$$
$$\vdots$$
$$f_n(t) = r_n \times \sin(\omega_n t)$$

But suppose we have subsequent frequencies that are integral multiples of the first one; then we can drop the subscript and simply use n to indicate an integer multiple, such as,

$$f_1(t) = r_1 \times \sin(\omega t)$$
$$f_2(t) = r_2 \times \sin(2\omega t)$$
$$f_3(t) = r_3 \times \sin(3\omega t)$$
$$f_4(t) = r_4 \times \sin(4\omega t)$$
$$\vdots$$
$$f_n(t) = r_n \times \sin(n\omega t)$$

Arbitrary Waveform

You may have noticed plotting data from a measurement does not usually follow a definite path of a curve, but it is not difficult to figure out if there is a repeated pattern in the data, and when one is present, this could be described as an arbitrary waveform. Even if there is no repeated pattern, we could still call it an arbitrary waveform, as we can argue that we didn't have enough time to monitor the event to see its repeated pattern. In any case, treat the data in hand as the data for one complete cycle. We are interested in describing the waveform mathematically in an algebraic form using trigonometric functions—in this description, the sine and cosine functions are the building blocks of the waveform—and this is what Fourier analysis is all about: giving an event a mathematical description as a function of frequency.

Converting a function into its Fourier series is a reversible process. If a function is decomposed into its component frequencies, adding these frequencies would bring back the original function. The decomposition is Fourier analysis and the recomposition is Fourier synthesis. In either case, we need to determine the amplitude, the period, and the total number of trigonometric functions that describe the waveform. It is easier to identify the period and the total number of trig functions (you will see this when we discuss Fourier synthesis), but the difficult part is identifying the amplitude of the sin and cosine waves and that will be deferred till the subsequent section of the Fourier analysis.

FOURIER SYNTHESIS

Since we are going to describe a function in terms of cyclic frequencies, we have to start from some fundamental frequency and build upon it.

A waveform that starts with 0 height may only have the sin components and not the cosine components; since the presence of cosine functions would force the waveform to start with a value other then 0, and that would defeat the purpose of synthesizing the desired waveform. Note that the period of one complete cycle of the waveform should be same as the period of the sine function. Otherwise, the ends won't meet. Thus, the first sine function has the period equal to the period of the arbitrary waveform. In other words, the frequency of the arbitrary waveform is the *fundamental frequency* of our first sine function.

With these assumptions, we denote the fundamental sin function as $a \sin(\omega t)$, where a is the amplitude or the *coefficient* (to be determined) of the sine wave and ω is the angular frequency equal to the frequency of the waveform.

A single sine function is not sufficient, so we add some more. The next function should have a different period but should still start at 0 and end at 0. This can be achieved only if the period of the next sine function is an *integer multiple* of the first one. Otherwise, we will end up with the same predicament in which the ends don't meet, and that would formulate a different function. (We will explore how to determine the total number of subsequent sine functions in the next section.)

We denote the fundamental sine wave and all subsequent sine waves as:

$$a_1 \times \sin(\omega t) + \\ a_2 \times \sin(2\omega t) + \\ a_3 \times \sin(3\omega t) + \\ \vdots \\ a_n \times \sin(n\omega t) \\ \stackrel{?}{=} \sum_{n=1}^{} a_n \times \sin(n\omega t)$$

For example, an arbitrary waveform function $f(t)$, as shown in Figure 1.7 (angular frequency $\omega = 2\pi \, radian/\sec$) may be synthesized by adding the following three sine functions

$$f(t) = 1.5 * \sin(2\pi t) + 0.5 * \sin(2 \times 2\pi t) + 0.3 * \sin(3 \times 2\pi t)$$

The other possibility of an arbitrary waveform is that it may begin with a nonzero height, indicating the presence of cosine waves in addition to the sine waves. As with the sine wave, the fundamental cosine function period must equal the period of the desired waveform. Also, the subsequent cosine function periods must be integer multiples of the fundamental periods. We denote these cosine waves as $b_n \times \cos(n\omega t)$ where b is the amplitude or the coefficient of the cosine wave, and ω is the frequency.

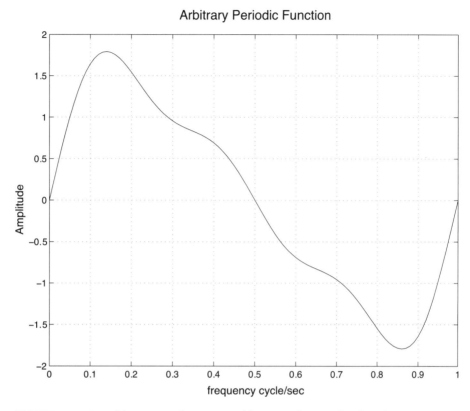

FIGURE 1.7 An arbitrary waveform created by summing up sine functions.

Using the sine and cosine functions, an arbitrary waveform may be described as,

$$f(t) =$$
$$a_1 \sin(\omega t) + b_1 \cos(\omega t) +$$
$$a_2 \sin(2\omega t) + b_2 \cos(2\omega t) +$$
$$a_2 \sin(3\omega t) + b_2 \cos(3\omega t) +$$
$$\vdots$$
$$a_n \times \sin(n\omega t) + b_n \times \cos(n\omega t)$$
$$= \sum_{n=1}^{?} a_n \sin(n\omega t) + \sum_{n=1}^{?} b_n \cos(n\omega t)$$

For example, adding the following two cosine functions to the arbitrary waveform of Figure 1.7, we see a new pattern emerging, as shown in Figure 1.8.

$$f(t) = 1.5\sin(2\pi t) + 0.5\sin(2 \times 2\pi t) + 0.3\sin(3 \times 2\pi t) + 2\cos(2\pi t) + 2\cos(3 \times 2\pi t)$$

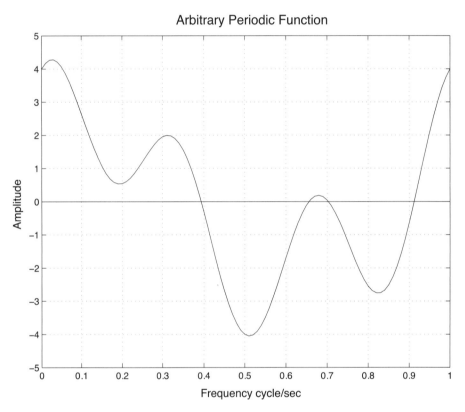

FIGURE 1.8 Addition of cosine waves to sine waves.

A complete cycle of a sine and cosine wave includes a positive valued portion and a negative valued portion. In the previous example, we saw how an arbitrary waveform was generated using only sine and cosine functions. If we try to find the total area under the curve, it should be equal to zero. In an arbitrary waveform, a nonzero value for an area of one complete cycle indicates the presence of a constant value in addition to having sine and cosine waves. You could see the effect of adding a constant value of 3.5 to the arbitrary waveform of the previous example, as shown in the Figure 1.9. The entire wave is almost shifted in the upper half of the axis. Now, we have a complete formula for an arbitrary function (Equation 1.1), there is a constant value (also called the *dc component* of the wave), and a series addition of sine and cosine functions.

$$f(t) = 3.5 + 1.5\sin(2\pi t) + 0.5\sin(2 \times 2\pi t) + 0.3\sin(3 \times 2\pi t) + 2\cos(2\pi t) + 2\cos(3 \times 2\pi t)$$
(1.1)

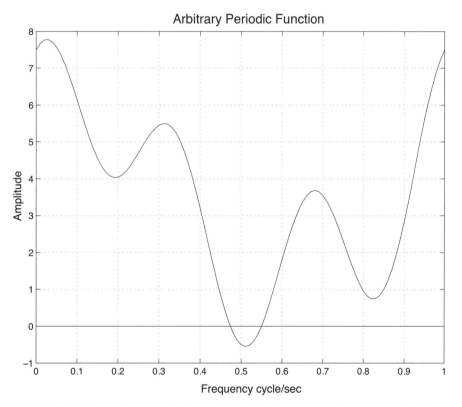

FIGURE 1.9 Addition of constant value of 3.5 shifts the wave to the upper half of the axis.

After adding the dc constant a_0, the arbitrary waveform $f(t)$ may now be described as,

$$f(t) = a_0 + \sum_{n=1}^{?} a_n \sin(n\omega t) + \sum_{n=1}^{?} b_n \cos(n\omega t)$$

16 Digital Signal Processing Fundamentals

This equation may also be described in terms of cyclic frequency as in Equation 1.2.

$$f(t) = a_0 + \sum_{n=1}^{?} a_n \sin(n2\pi ft) + \sum_{n=1}^{?} b_n \cos(n2\pi ft) \tag{1.2}$$

Or if the period of one complete cycle is given as T, Equation 1.2 becomes

$$f(t) = a_0 + \sum_{n=1}^{?} a_n \sin(\frac{n2\pi}{T}t) + \sum_{n=1}^{?} b_n \cos(\frac{n2\pi}{T}t)$$

The two things we have determined so far in composing an arbitrary waveform are that the fundamental frequency is the same as the frequency of the arbitrary waveform, and the subsequent frequencies are integer multiples of the fundamental frequency. The remaining two aspects are determining the amplitudes and the total number of waves.

FOURIER ANALYSIS

In this section, we will analyze an arbitrary waveform and determine its constituent frequencies.

Given an arbitrary function the task is:

- To find the starting or *fundamental* frequency of the series that made the waveform
- To find the total number of waves in the series
- To find the coefficients or the amplitudes of the sine and cosine waves in the series
- To find the dc component of the series

EXAMPLE 1.1

We proceed with the analysis of a physical event by using data acquired by a digital computer. The data acquisition rate was 16 samples per second and the data for one complete cycle is presented in Table 1.1. The graphical representation of data is shown in Figure 1.10. We would like to analyze this function to find its component frequencies as described by Fourier.

TABLE 1.1 Data Acquisition Values of 16 Events, at Intervals of 0.0625 Seconds Apart

K	kΔt	2πfkΔt	f(t)	K	kΔt	2πfkΔt	f(t)
0	0	0	7.5	8	0.5	3.2	-0.55
1	0.0625	0.4	7.29	9	0.5625	3.6	0.76
2	0.125	0.8	5.2	10	0.625	4	3.08
3	0.1875	1.2	4.03	11	0.6875	4.4	3.54
4	0.25	1.6	4.79	12	0.75	4.8	1.86
5	0.3125	2	5.49	13	0.8125	5.2	0.74
6	0.375	2.4	4	14	0.875	5.6	2.43
7	0.4375	2.8	1.02	15	0.9375	6	5.83

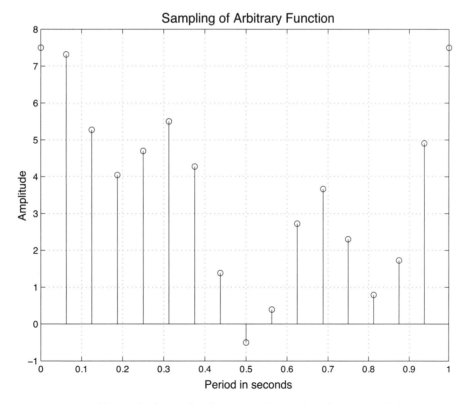

FIGURE 1.10 This graph shows the data acquisition values for one period.

Fundamental Frequency

If T is the period of the arbitrary waveform and the sampling period is Δt, then the number of samples $K = T/\Delta t$. We could define the fundamental frequency either in terms of number of samples N or the period $1/T$. Thus, the fundamental sine frequency is $a_1 \sin(2\pi k\Delta t/T)$, and the fundamental cosine frequency is $b_1 \cos(2\pi k\Delta t/T)$.

From the data given in Table 1.1, a total of 16 samples were taken during 1 sec period, thus the fundamental frequency $f = 1$ sec.

Since all other frequencies must be integer multiples of the fundamental frequency, the rest may be described as $a_n \sin(n 2\pi k\Delta t/T)$ and $b_n \cos(n 2\pi k\Delta t/T)$. The discrete time version of Equation 1.2 may be given as,

$$f(k) =$$
$$a_0 +$$
$$a_1 \sin(2\pi k\Delta t/T) + b_1 \cos(2\pi k\Delta t/T) +$$
$$a_2 \sin(4\pi k\Delta t/T) + b_2 \cos(4\pi k\Delta t/T) +$$
$$a_2 \sin(6\pi k\Delta t/T) + b_2 \cos(6\pi k\Delta t/T) +$$
$$\vdots$$
$$a_n \sin(n 2\pi k\Delta t/T) + b_n \cos(n 2\pi k\Delta t/T)$$
$$= a_0 + \sum_{n=1}^{?} a_n \sin(n 2\pi k\Delta t/T) + \sum_{n=1}^{?} b_n \cos(n 2\pi k\Delta t/T)$$

The previous example had 16 samples for one complete set of information that makes the fundamental period $T = 1$sec and the sampling interval $\Delta t = 1/16$ or 0.0625 seconds per sample. The k is an integer and has units of "sample."

$$k = 0..15$$
$$f = 1 cycle / \sec$$
$$\Delta t = 1/16 = 0.0625 \sec$$
$$T = 1 \sec$$
$$2\pi k \Delta t / T = 0.4k$$

The fundamental constituent frequencies are

$$a_1 \sin(0.4k), \; b_1 \cos(0.4k)$$

We begin with the fundamental frequency and describe the discrete Fourier series of the example arbitrary waveform as in Equation 1.3.

$$f(k) = a_0 + \sum_{n=1}^{?} a_n \times \sin(n0.4k) + \sum_{n=1}^{?} b_n \times \cos(n0.4k) \qquad (1.3)$$

The coefficients $a0$, an, bn, and the total number of frequencies in Equation 1.3 are still to be determined.

Harmonics

Harmonics are the integer multiples of the fundamental frequency, but the total number of frequencies in a system is different for the discrete time analysis and for the continuous time analysis. You may have noticed in the section on Fourier synthesis that as we started adding more frequencies to the series, the picture of the original waveform started getting clearer. Ideally, for a continuous time function, it may require infinite frequencies to formulate an arbitrary waveform, especially if there are jump discontinuities in the function; this is Gibb's phenomenon and we will discuss it later. But for a discrete time function, there is a limit to the number of frequencies that can be identified.

The Number of Frequencies in the Arbitrary Waveform

In discrete time processing, the time t is valid only at the instance $k\Delta t$ when the sample is being acquired. It is important for the sample points to repeat the pattern in the subsequent cycles for the solution to be valid for periodic functions. If there are frequencies in the original waveform whose period is shorter then the sampling period Δt, it is quite possible that our sampling process would pick up these frequencies at the wrong time. The result would be totally unpredictable, just like the old western movies where wheels turn backward and the stagecoach moves forward: the camera could not take pictures fast enough, because the number of spokes of the wheel passing through one point in one second was faster then the number of frames grabbed by the camera in one second.

To see the effect in real life, imagine there is only one frequency of 4 Hz in the input signal. If the sampling rate is 3 Hz and we use that to reconstitute the original wave, it would give us the answer of 6 Hz, a totally wrong answer for the original wave approximation.

We are essentially introducing *alias frequencies* into our system by not being able to process data fast enough. In practice, input frequencies are filtered out using hardware techniques before analog-to-digital conversion has taken place. There is a limit to the maximum frequency that a data acquisition system is capable of processing.

Nyquist Frequency

Nyquist frequency is the limit beyond which we start introducing alias frequencies into our system. It is clear that a slow sampling rate may distort the reconstruction of the desired waveform using a Fourier series. We need two samples per period to determine at least one period of the wave. This makes the number of frequencies in a waveform equal to half the number of samples in one cycle. Likewise, the last frequency that can be recognized is half the sampling rate of the system. The last frequency is also called the Nyquist frequency of the system for a given sampling rate.

Referring back to Example 1.1, the sampling rate of 16 samples per second limits us to the maximum input frequency of 8 Hz. With this limitation, we would reintroduce the Fourier series summation process of Equation 1.3 with the number of frequencies $N = 8$: frequencies 0..7 (Equation 1.4).

$$f(t) = a_0 + \sum_{n=1}^{N-1} a_n \times \sin(n0.4k) + \sum_{n=1}^{N-1} b_n \times \cos(n0.4k) \tag{1.4}$$

So far we have managed to complete two out of the four tasks for our example problem: the starting (fundamental) frequency is 1 Hz, and the total number of frequencies in the series $N = 8$. Next we must find the amplitude.

Amplitude of the Sine and Cosine Waves

Now we discuss the difficult part: finding the amplitude of the individual component waves. The method is so brilliant that you cannot help but appreciate the genius in Fourier. His method has opened up a new branch of mathematics, and in honor of his contribution, the process is called finding *Fourier Coefficients* instead of finding the amplitude of the waves. The amplitude of a sine wave is a Fourier sin coefficient, and the amplitude of a cosine wave is a Fourier cosine coefficient. The method is based on the orthogonal property of the trigonometric functions.

Orthogonal Functions

A complete cycle of sine and cosine functions includes a positive valued portion and a negative valued portion. The result of multiplying a sine function with another sine function would produce equally positive and negative values, but when a sine function is multiplied by a sine function with the same frequency or a cosine function is multiplied by another cosine function with the same frequency there would be no negative value as shown in Figure 1.11. If the result of multiplying a sine function by any other sine function is integrated for one complete cycle, it should add up to be equal to zero, except when the frequency matches with the function itself. (Simply adding the little rectangles (see Figure 1.12) formed by multiplying the height and the length Δt could complete the integration). Orthogonal functions are those that when multiplied and integrated produce zero area, except when they are multiplied by themselves; trigonometric functions fall into this category.

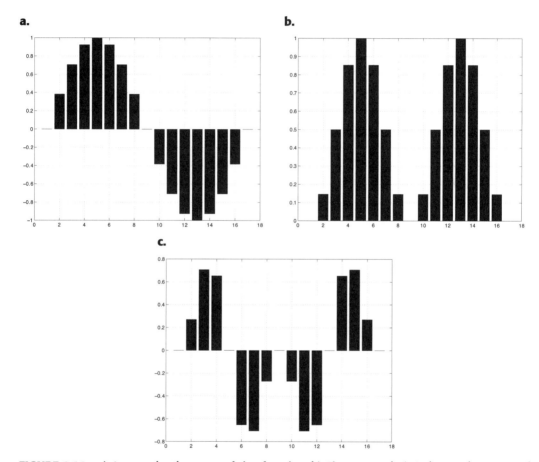

FIGURE 1.11 a) Area under the curve of sine function. b) Sine squared. c) Orthogonal property of the sine function.

Figure 1.11 shows the orthogonal property Fourier exploited to extract the sine and cosine coefficients for the trigonometric series of the arbitrary waveform.

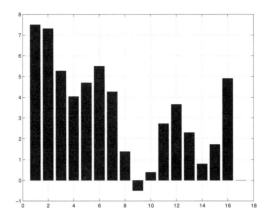

FIGURE 1.12 Integrating the function using the rectangular area method.

Computing Fourier Coefficients a_n and b_n

The square integral of the trigonometric function acts like a filter of frequency where only the matching frequencies survive. You can see this graphically in Figure 1.11. The function $a_n \sin(\omega t)$ is multiplied by $\sin(\omega t)$ and integrated for one complete cycle of period T. The area produced by the multiplication and summation is equal to half the area of the rectangle formed *height* × *width*,

$$Area = \frac{a_n \times T}{2}$$

$$\frac{a_n \times T}{2} = \sum_{k=0}^{K-1} f(k) \times a_n \sin(n2\pi k \Delta t / T) \times \Delta t$$

Thus, from the result of summing all the areas, we could obtain the coefficient a_n as,

$$a_n = \frac{2}{T} \sum_{k=0}^{K-1} f(k) \times a_n \sin(n2\pi k \Delta t / T) \times \Delta t$$

Similarly, the cosine coefficients for discrete time processing are obtained as,

$$b_n = \frac{2}{T} \sum_{k=0}^{K-1} f(k) \times a_n \cos(n2\pi k \Delta t / T) \times \Delta t$$

The process should be repeated for all the frequencies in the waveform up to the Nyquist frequency in the system for discrete time processing and up to infinity for continuous time processing.

Computing the dc Constant a_0

We know the effect of the dc constant a_0: it shifts the entire wave either below or above the base line. Suppose we did not have the dc constant; the area under the curve of the complex wave would be exactly equal to 0 over one complete period. Any nonzero value for the area indicates contribution of the dc component to the complex wave. To find the area for one period T, we can simply add all the little rectangles we get by multiplying each sample point $f_k(k)$ with ΔT of sampling frequency, as shown in Figure 1.9. The constant a_0 is obtained by dividing the area by the width, as in:

$$Height = Area/Width$$

$$a_0 = \frac{\sum_{k=0}^{K-1} f_k(k) \times \Delta T}{T}$$

This gives us a complete picture of Fourier analysis by finding the constituent frequencies in an arbitrary waveform.

Discrete Time Fourier Series

The Fourier series of an arbitrary waveform obtained with discrete time sampling is described in Equation 1.5.

$$f(k) = a_0 + \sum_{n=1}^{N-1} a_n \times \sin(n2\pi k \Delta t/T) + \sum_{n=1}^{N-1} b_n \times \cos(n2\pi k \Delta t/T) \tag{1.5}$$

In terms of *Nyquist frequency N*, see Equation 1.6.

$$f(k) = a_0 + \sum_{n=1}^{N-1} a_n \times \sin(n2\pi k/K) + \sum_{n=1}^{N-1} b_n \times \cos(n2\pi k/K) \tag{1.6}$$

In this equation:

N: The Nyquist frequency (Total number of samples / 2)
k: The sample number
T: The period of the arbitrary waveform
K: Total number of samples

Equation 1.7 shows the dc Constant:

$$a_0 = \frac{1}{T}\sum_{k=0}^{K-1} f_k(k) \times \Delta T \qquad (1.7)$$

Equation 1.8 shows the sine coefficients:

$$a_n = \frac{2}{T}\sum_{k=0}^{K-1} f(k) \times a_n \sin(n 2\pi k \Delta t/T) \times \Delta t \qquad (1.8)$$

Equation 1.9 shows the cosine coefficients:

$$b_n = \frac{2}{T}\sum_{k=0}^{K-1} f(k) \times a_n \cos(n 2\pi k \Delta t/T) \times \Delta t \qquad (1.9)$$

Referring back to Example 1.1, Table 1.2 shows the computation values for the coefficients of the fundamental frequency.

The *fundamental harmonics* provided in the last column of Table 1.2 is $a1 = 1.5$. The following coefficients for the rest of the harmonics $n=1..7$, were obtained with the similar method and are tabulated as follows,

Fourier sine and cosine coefficients:

$$a_1 = 1.5........b_1 = 2$$
$$a_2 = 0.5........b_2 = 0$$
$$a_3 = 0.3........b_3 = 2$$
$$a_4 = 0..........b_4 = 0$$
$$a_5 = 0..........b_5 = 0$$
$$a_6 = 0..........b_6 = 0$$
$$a_7 = 0..........b_7 = 0$$

We get the following series by substituting the Fourier coefficients into the Fourier series formula of Equation 1.3

$$a_1 = 1.5........b_1 = 2$$
$$a_2 = 0.5........b_2 = 0$$
$$a_3 = 0.3........b_3 = 2$$
$$a_4 = 0..........b_4 = 0$$

TABLE 1.2. Computation Values for the Coefficients of the Fundamental Frequency.

k	Function value $f(k)$	Sine multiplier $f(k) \times \sin(2\pi k \Delta T / T)$	Area of rectangle $f(k) \times \sin(2\pi k \Delta T / T) \Delta T$
0	7.5	0	0
1	7.32	2.8	0.175
2	5.27	3.73	0.233
3	4.04	3.73	0.233
4	4.7	4.7	0.294
5	5.5	5.08	0.318
6	4.27	3.02	0.189
7	1.38	0.53	0.033
8	−0.5	0	0
9	0.39	−0.15	−0.01
10	2.73	−1.93	−0.12
11	3.66	−3.39	−0.212
12	2.3	−2.3	−0.144
13	0.79	−0.73	−0.045
14	1.73	−1.22	−0.076
15	4.91	−1.88	−0.117
		$a_1 = \dfrac{2}{T} \sum_{k=0}^{K-1} f(k) \times \sin(2\pi k \Delta T / T) \Delta T$	1.5

$$a_5 = 0 \ldots\ldots\ldots b_5 = 0$$
$$a_6 = 0 \ldots\ldots\ldots b_6 = 0$$
$$a_7 = 0 \ldots\ldots\ldots b_7 = 0$$

The dc Constant a_0

The constant a_0 is obtained by dividing the total area with the total width.

$$Height = Area/Width$$

$$a_0 = \frac{\sum_{k=0}^{K-1} f_k(k\Delta T) \times \Delta T}{T} = 3.5$$

The Fourier series for the arbitrary waveform described by the data points of Example 1.1 may be obtained by substituting the Fourier coefficients and the harmonics as shown in Equation 1.10

$$f(k) = 3.5 + 1.5\sin(0.4k) + 0.5\sin(0.8k) + 0.3\sin(1.2k) + 2\cos(0.4k) + 2\cos(1.2k) \quad (1.10)$$

Equation 1.10 is the same as Equation 1.1 when we substitute for $k = 1..16$.

The process of Fourier analysis describes a function as a sum of series of sine and cosine frequencies, and once the component frequencies are identified, we may perform further analysis of the system as if each were a separate input to the system.

Continuous Time Fourier Series

We developed the discrete time version of the Fourier series (as given in Equation 1.6) with the help of a function whose values were acquired at discrete interval of Δt, and the data for one complete cycle was obtained in time period T. Next, we would like to develop the continuous time version of the Fourier series, and for that, we need some help from calculus. (You may skip the next section if you are not familiar with the calculus terminology of limit and integration.)

We begin with rewriting the Fourier series of Equation 1.2 by filling in the upper limit of summation with infinity, as there is no end to the number of frequencies in continuous time (Equation 1.11).

$$f(t) = a_0 + \sum_{n=1}^{\infty} a_n \times \sin(n 2\pi f t) + \sum_{n=1}^{\infty} b_n \times \cos(n 2\pi f t) \quad (1.11)$$

We computed the Fourier coefficients using the orthogonal property of the sine and cosine functions for the discrete time interval Δt (see Equations 1.7, 1.8, and 1.9). If we were to shorten the time delta of the sampling period to dt and take it to the limit as dt approaches 0, the summation process in Fourier coefficients would become an integration, and the summation limit would be from 0 to 2π, indicating integration over one complete cycle of 2π radians.

The area produced by the multiplication and integration is equal to half the area of the rectangle formed by $height = width$,

$$Area = \frac{a_n \times 2\pi}{2}$$

And the coefficient a_n is shown in Equation 1.12.

$$a_n = \frac{1}{\pi} \int_{t=0}^{2\pi} f(t) a_n \sin(\omega_n t) dt \qquad (1.12)$$

The cosine coefficients for continuous time processing are shown in Equation 1.13.

$$b_n = \frac{1}{\pi} \int_{t=0}^{2\pi} f(t) b_n \cos(\omega_n t) dt \qquad (1.13)$$

The dc constant is shown in Equation 1.14.

$$a_0 = \frac{1}{2\pi} \int_0^{2\pi} f(t) dt \qquad (1.14)$$

We conclude the discussion of Fourier analysis with an example problem of a periodic disturbance in a physical system where Fourier analysis provides a logical solution.

EXAMPLE 1.2

A periodic force of 20 lb magnitude is applied on a system comprising a spring, mass, and a dashpot, as shown in Figure 1.13. Not knowing that hidden in the input force there are some frequencies that are close to the resonant or natural frequency of the system, a mysterious response will appear. The system will start oscillating with a frequency of 2 cycles per second while the input is only 1 cycle per second. We could predict this response of the system with the Fourier analysis of the input source by identifying the component frequencies that are close to the resonant or natural frequency of the system.

The solution presented here requires knowledge of differential equations, which will be covered in Chapters 2, "Complex Number Arithmetic" and 4, "Solutions of Differential Equations." If you are not familiar with the terminology, you may revisit this part once you have covered the subsequent chapters.

First, identify the component frequencies present in the input force through Fourier analysis.

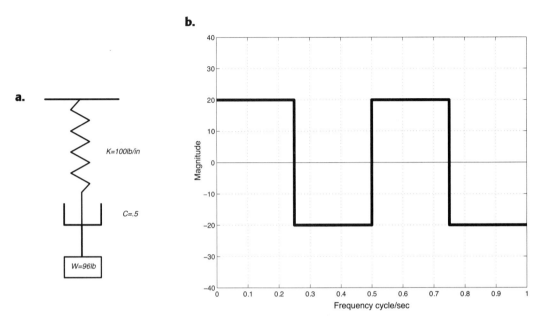

FIGURE 1.13 a) A spring, mass, and dashpot system. b) Integrating the function using the rectangular area method.

Computing the Fourier Coefficients for the Driving Force

The dc constant may be computed using Equation 1.7, but a glance at the driving force tells us that the upper halves and lower halves are equal, thus there is no dc component in the input force,

$$a_0 = 0$$

The cosine coefficients may be computed using Equation 1.9, but the function that begins at 0 and ends at 0 tells us there is no cosine component in the driving force,

$$b_n = 0$$

The sine coefficients are computed using Equation 1.8, but the integral would cancel the positive area and the negative area. We can avoid this by integrating only half the area and multiplying it by 2, resulting in Equation 1.15.

$$a_n = \frac{2}{1/2} \int_0^{1/2} 20\sin(\frac{n\pi t}{1/2})dt$$

$$a_n = 80\left[-\frac{\cos 2n\pi t}{2n\pi}\right]_0^{1/2}$$

$$a_n = 40 \times \frac{1-\cos n\pi}{n\pi} \qquad (1.15)$$

For every even n in Equation 1.10 the coefficient value of $b_n = 0$

$$a_n = 0 \quad n \text{ even}$$

For odd ns, we get the following coefficient

$$a_n = \frac{80}{n\pi} \quad n \text{ odd}$$

Thus the component frequencies in the driving force are:

$$F(t) = \frac{80}{\pi}\left[\sin 2\pi t + \frac{\sin 6\pi t}{3} + \frac{\sin 10\pi t}{5} + \frac{\sin 14\pi t}{7} + \frac{\sin 18\pi t}{9} \cdots\right]$$

Each component frequency in the Fourier series contributes independently to the outcome of the system. For the sake of brevity, we will compute the magnitude of the forces contributed by the first few terms only, as in Equation 1.16.

$$F_0(t) = \frac{80}{\pi}, F_1(t) = \frac{80}{3\pi}, F_2(t) = \frac{80}{5\pi}, F_3(t) = \frac{80}{7\pi}, F_4(t) = \frac{80}{9\pi}, \qquad (1.16)$$

Systems Response

The systems response is the natural response plus the forced response. We will analyze the natural response of the spring, mass, and dashpot system of Figure 1.13 with the help of Newton's law:

$$Force = mass \times acceleraton$$

We are interested in finding out if there is any component frequency in the input force that matches the natural response frequency of the system, and if there is one, we can predict that the system will oscillate to the natural frequency with an amplified response. The total response will be the sum of the natural response and the forced response.

Assuming only the weight w of the mass m is significant and the acceleration due to vibration is only in the y direction, the natural response could be obtained as follows,

$$mass = w/g \qquad\qquad acceleration = \frac{d^2y}{dt^2}$$

The force due to gravity is

$$Fg = -w$$

The force with which the spring exerts upon the weight is

$$Fk = w - ky$$

The force due to viscous damping is

$$Fc = -c\frac{dy}{dt}$$

Substituting all the forces into Newton's law, we get

$$\frac{w}{g}\frac{d^2y}{dt^2} = -w + (w - ky) - c\frac{dy}{dt}$$

After rearranging and summing all the forces, we obtain the natural response as in Equation 1.17.

$$\frac{w}{g}\frac{d^2y}{dt^2} + c\frac{dy}{dt} + ky = 0 \qquad(1.17)$$

The forces applied to the system of Equation 1.17 are given in the Fourier series of Equation 1.16. We could find the complete solution by computing the response due to each individual force and adding them all later, but there is a better solution.

If we change Equation 1.17 into a frequency dependent function (known as a *transfer function*), by simply plugging in the frequencies and magnitudes of the Fourier series of Equation 1.16, we could analyze the system response. This method is discussed in detail in Chapters 4, "Solutions of Differential Equations," and 5, "Laplace Transforms and z-Transform," but for now, we will proceed with the characteristic solution of the Equation 1.17, as follows.

The characteristic equation is shown in Equation 1.18.

$$\frac{w}{g}s^2 + cs + k = F_0 \tag{1.18}$$

Substituting the value for $s = j\omega$ in Equation 1.18, we get the output as complex valued function,

$$Y = \frac{F_0}{\frac{w}{g}(j\omega)^2 + cj\omega + k}$$

The magnitude as a function of frequency is obtained by separating real and imaginary components, as follows

$$Y = \frac{F_0}{\sqrt{(k - \frac{w}{g}\omega^2)^2 + (c\omega)^2}}$$

And the phase response is

$$\phi = \tan^{-1}\frac{c\omega}{k - \frac{w}{g}\omega^2}$$

For each input force and the frequency as defined in Equation 1.16, we could tabulate the output as shown in Table 1.3.

The last column of Table 1.3 provides the response due to individual frequency components present in the input force applied to the system. A comparison of the magnitudes of the output response would indicate a dominant second harmonics in the input force (magnitude 0.58), and that explains why the example system started oscillating with twice the frequency of the input force, as shown in the Figure 1.14.

TABLE 1.3 Tabulation of the Response of Individual Input Frequencies for the System of Example 1.1

Force applied	ω	$\dfrac{1}{\sqrt{(k-\dfrac{w}{g}\omega^2)^2+(c\omega)^2}}$	Phase $\phi=\tan^{-1}\dfrac{c\omega}{k-\dfrac{w}{g}\omega^2}$	Response
$F_0 = \dfrac{80}{\pi}\sin\dfrac{(2\pi t)}{7}$	2π	.011	2	$0.28\sin(2\pi t - 2)$
$F_0 = \dfrac{80}{\pi}\sin\dfrac{(2\pi t)}{7}$	6π	.048	40	$0.58\sin(6\pi t - 40)$
$F_5 = \dfrac{80}{5\pi}\sin\dfrac{(2\pi t)}{5}$	10π	.006	174	$0.03\sin(10\pi t - 17)$
$F_7 = \dfrac{80}{7\pi}\sin\dfrac{(2\pi t)}{7}$	14π	.002	177	$0.01\sin(14\pi t - 17)$

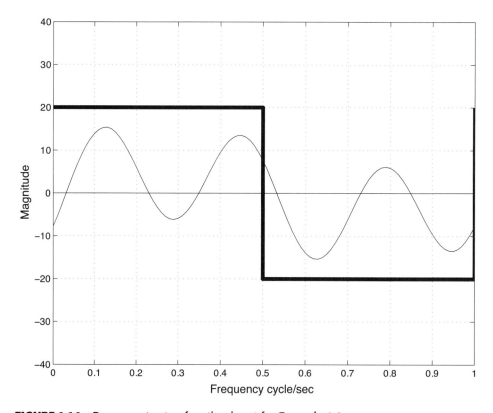

FIGURE 1.14 Response to step function input for Example 1.2.

SUMMARY

We have studied Fourier analysis as a method of describing an arbitrary function by its frequency components, a sum of series of sine and cosine functions with a dc component. The technique provides us with the capability of analyzing a physical system using a mathematical description. We have shown with an example that Fourier analysis can predict the response of a system that otherwise might be very difficult to obtain in time domain.

A continuous time Fourier series of a function with the period $T = \frac{2\pi}{\omega}$ may be described as:

$$f(t) = a_0 + \sum_{n=1}^{\infty} a_n \times \sin(n\omega t) + \sum_{n=1}^{\infty} b_n \times \cos(n\omega t)$$

The dc constant:

$$a_0 = \frac{1}{2\pi} \int_0^{2\pi} f(t)dt$$

The sine coefficients:

$$a_n = \frac{1}{\pi} \int_{t=0}^{2\pi} f(t) a_n \sin(\omega_n t) dt$$

The cosine coefficients:

$$b_n = \frac{1}{\pi} \int_{t=0}^{2\pi} f(t) b_n \cos(\omega_n t) dt$$

A discrete time Fourier series of a function with the Nyquist frequency N and the period T may be described as:

$$f(k) = a_0 + \sum_{n=1}^{N-1} a_n \times \sin(n 2\pi k \Delta t / T) + \sum_{n=1}^{N-1} b_n \times \cos(n 2\pi k \Delta t / T)$$

or:

$$f(k) = a_0 + \sum_{n=1}^{N-1} a_n \times \sin(n\pi k/(2N)) + \sum_{n=1}^{N-1} b_n \times \cos(n\pi k/(2N))$$

The dc constant:

$$a_0 = \frac{1}{T} \sum_{k=0}^{K-1} f_k(k) \times \Delta T$$

The sine coefficients:

$$a_n = \frac{2}{T} \sum_{k=0}^{K-1} f(k) \times a_n \sin(n 2\pi k \Delta t/T) \times \Delta t$$

The cosine coefficients:

$$b_n = \frac{2}{T} \sum_{k=0}^{K-1} f(k) \times a_n \cos(n 2\pi k \Delta t/T) \times \Delta t$$

2 Complex Number Arithmetic

In This Chapter

- Complex Number Representation
- Complex Numbers in Polar Coordinates
- The Exponent *e* and the Power Functions
- The Phasor Method
- Signal Modulation

We established in the previous chapter on Fourier analysis that an arbitrary function could be described in terms of simple sine and cosine functions. The mathematics of the sinusoid was developed using the example of a spot on a bicycle wheel moving in time. We could also consider the amplitude of the sinusoidal wave as if it was a point on a circle. We needed two numbers to describe the point, in a Cartesian coordinate system, the point $P(x, y)$ is the distance in the x direction, and the distance in the y direction, while in a polar coordinate system, was $r_n\cos(\theta)$ and $r_n\sin(\theta)$. But despite the fact that they were real numbers, we could not perform normal algebraic operations of addition and multiplication upon them. Point (x, y) in essence is a single number and is treated as one entity. For one, we could not perform arithmetic because the algebra does not allow us to have a comma in parentheses, so mathematicians had to invent a different numbering system, just like they did for negative numbers, real numbers, and logarithmic numbers. For representing points in a coordinate system, they invented *complex numbers*.

In this chapter, you'll see how the new numbering system of complex numbers allows us to apply algebraic rules when such numbers are placed in algebraic equations. We will develop the mathematical foundation and establish the rules of arithmetic operations involving complex numbers that will not only help us analyze the problems in electrical circuits (the same electrical circuits that we intend to simulate using digital signal processing) but also lay down the foundation for solving a broad range of problems that digital signal processing is intended to solve. We will perform filtering of frequencies, besides integration, differentiation, and smoothing of sampled data obtained from the analog world.

COMPLEX NUMBER REPRESENTATION

If you know Pythagoras theorem well, you will have no difficulty in understanding the complex number arithmetic. As mentioned before, a complex number is nothing but a point on a coordinate system made up of two numbers, essentially, the relative magnitude of the distance in x and y direction from the origin, whereas the absolute distance is computed as the square root of x squared and y squared (according to the Pythagoras theorem of the right angle triangle). Figure 2.1 shows the points in a Cartesian coordinate system.

Suppose you have two points P1 (1,3) and P2 (3,2) and you would like to add them together, the result is the third point P3 (4,5), as shown in Equation 2.1.

$$(1,3) + (3,2) = (4,5) \qquad (2.1)$$

Equation 2.1 is not algebraically correct: commas are not allowed in parentheses. So how do you represent two numbers as distinct as x and y and be algebraically correct? For this purpose, Carl Friedrich Gauss (1777–1855, Brunswick, Germany) introduced the concept of an *imaginary operator* $\sqrt{-1}$. If you multiply the second number by $\sqrt{-1}$, you can treat the x and y components of a complex number as if they were two ordinary real numbers as in

$$(1 + \sqrt{-1}(3)) + (3 + \sqrt{-1}(2)) = (4 + \sqrt{-1}(5))$$

It is customary to show the coordinate system y-axis as the imaginary axis and the x-axis as the real axis.

The Imaginary Operator $\sqrt{-1}$

The Gaussian operator $\sqrt{-1}$ helps us perform vector arithmetic using simple algebraic rules. Multiply the y component of the complex number by $\sqrt{-1}$ and

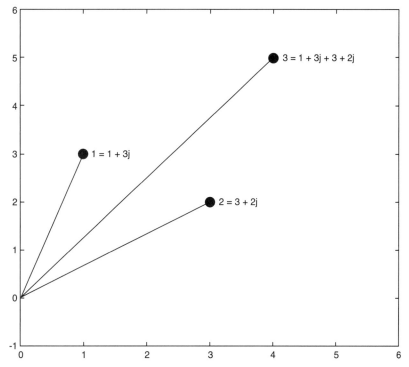

FIGURE 2.1 Points in Cartesian coordinate system.

all arithmetic operation can be carried out as ordinary algebraic quantities, as in

$$(a+\sqrt{-1}c) \times (b+\sqrt{-1}d) = (ab-cd) + \sqrt{-1}(bc+ad)$$
$$(a+\sqrt{-1}c) + (b+\sqrt{-1}d) = (a+b) + \sqrt{-1}(c+d)$$
$$(a+\sqrt{-1}c) - (b+\sqrt{-1}d) = (a-b) + \sqrt{-1}(c-d)$$
$$\frac{(a+\sqrt{-1}c)}{(b+\sqrt{-1}d)} = \frac{(a+\sqrt{-1}c)}{(b+\sqrt{-1}d)} \times \frac{(b-\sqrt{-1}d)}{(b-\sqrt{-1}d)} = \frac{(ab+cd) + \sqrt{-1}(bc-ad)}{(b^2+d^2)}$$

The operator $\sqrt{-1}$ (usually written as j or i in an equation) follows the rules of multiplication.

$$j = \sqrt{-1}, j^2 = -1, j^3 = -\sqrt{-1}, j^4 = 1, j^5 = \sqrt{-1}.....$$

Complex Conjugate

If $z = x + jy$ is an arbitrary complex number, $z = x - jy$ is the mirror image on the y-axis. The number $z = x - jy$ is the complex conjugate of $z = x + jy$, as shown in Figure 2.2b.

The complex conjugate is usually denoted with an asterisk

$$z^* = x - jy$$

The conjugate pair of complex numbers has the following property,

$$\frac{1}{2}(z + z^*) = \frac{1}{2}\left[(x + jy) + (x - jy)\right] = x = \text{Re}(z)$$

$$\frac{1}{2}(z - z^*) = \frac{1}{2}\left[(x + jy) - (x - jy)\right] = jy = j\,\text{Im}(z)$$

The square magnitude of the complex number is obtained by multiplying it by its complex conjugate, as in

$$zz^* = \left[(x + jy) \times (x - jy)\right] = (x^2 + y^2)$$

Thus,

$$|4 + j3| = \sqrt{(4 + j3) \times (4 - j3)} = \sqrt{(4)^2 + (3)^2} = 5$$

To add the two points $(3 + 2j)$ and $(4 + 5j)$, add the real values and the imaginary values separately. The result is a new complex number, shown graphically in Figure 2.2.

$$(3 + 2j) + (4 + 5j) = 7 + 7j$$

Multiplying a real value by j produces an imaginary number, and multiplying an imaginary number by j results in a real value but in the negative direction. For example, multiplying the real value 5 with j produces $j5$, which is on the imaginary axis; subsequent multiplication of $j5$ with j results in -5, which is on real axis in the negative direction; another multiplying of j with -5 produces $-j5$, which is on negative imaginary axis; subsequent multiplying of $-j5$ with j returns back to the real value 5. Figure 2.3 shows how the repeated multiplication with j rotates the number in a counterclockwise direction. Similarly, multiplication with $-j$ results in a clockwise rotation of 90° on the complex plane.

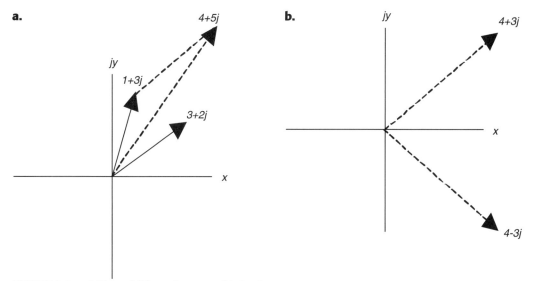

FIGURE 2.2 a) The addition of vectors. b) Conjugate vectors.

A *complex plane* is a Cartesian coordinate system in which the distance along the vertical axis is measured in units of $j\,(j=\sqrt{-1})$, and those along the horizontal axis in the usual units of 1, as shown in Figure 2.4. The y-axis is the imaginary axis and the x-axis is the real axis.

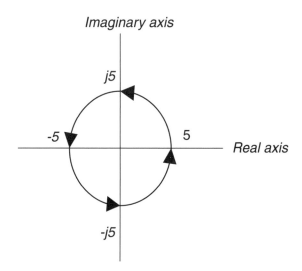

FIGURE 2.3 Multiplying with j rotates the vector to 90° counterclockwise.

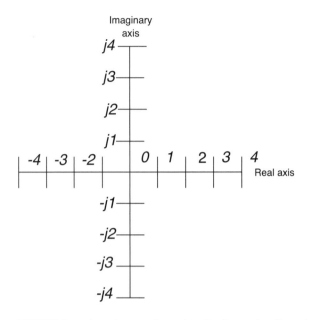

FIGURE 2.4 Imaginary axis and real axis on the Cartesian coordinate system.

COMPLEX NUMBERS IN POLAR COORDINATES

The position of the complex numbers in Cartesian coordinates determines the distance in the x and y directions. Sometimes it is useful to represent the complex number in polar coordinates, where the complex number is represented as a directed line OP (a vector of length M) rotated from the initial line OA through an angle θ. Figure 2.5 shows three complex numbers, OP1 rotated at an angle of 70°, OP2 rotated at an angle of 30°, and OP3 rotated at an angle of 50°. The complex number in polar form with magnitude vector M at angle θ may be written as

$$P = M \angle \vartheta \qquad (2.2)$$

To convert complex numbers in polar coordinates to rectangular form,

$$x = M \cos \vartheta$$
$$y = M \sin \vartheta$$

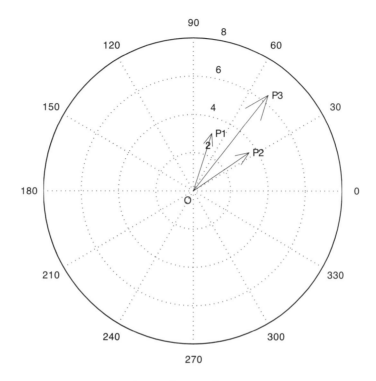

FIGURE 2.5 Vectors in a polar coordinate system.

We can convert the values from the Cartesian coordinates to the polar coordinates using the trigonometric rule

$$r = \sqrt{x^2 + y^2}, \theta = \tan^{-1}\frac{x}{y}$$
$$x = r\cos(\theta), y = r\sin(\theta)$$

The polar form of the complex number is

$$p = r(\cos(x) + j\sin(x))$$

If the point is on a unit radius circle (taking r as a unit magnitude), we can write the number as

$$p = (\cos(x) + j\sin(x))$$

While performing addition and subtraction of complex numbers is easier in rectangular coordinates, there is a definite advantage in doing multiplication and division in polar coordinates. The rules are simple: for multiplication, multiply the magnitude and add the angles, and for division, divide the magnitude and subtract the angles, as in

$$z_1 = r_1(\cos\theta_1 + j\sin\theta_1)$$

and

$$z_2 = r_2(\cos\theta_2 + j\sin\theta_2)$$

Then,

$$z_1 z_2 = r_1 r_2 \left[\cos(\theta_1 + \theta_2) + j\sin(\theta_1 + \theta_2)\right]$$

And

$$\frac{z_1}{z_2} = \frac{r_1}{r_2}\left[\cos(\theta_1 - \theta_2) + j\sin(\theta_1 - \theta_2)\right]$$

The multiplication and division can also be performed symbolically,

$$(P\angle\alpha)(Q\angle\beta) = PQ\angle\alpha+\beta$$

$$\frac{(P\angle\alpha)}{(Q\angle\beta)} = \frac{P}{Q}\angle\alpha-\beta$$

$$(P\angle\alpha)^n = P^n\angle n\alpha$$

$$(P\angle\alpha)^{1/n} = P^{1/n}\angle\alpha/n$$

The angle θ, also referred to as the *phase* angle, indicates the rotation of the magnitude vector M from the real axis, as shown in Figure 2.5.

Powers of Complex Numbers

To get the square of a complex number, we can use the trigonometric identity,

$$[r(\cos\theta + j\sin\theta)]^2 = r^2(\cos 2\theta + j\sin 2\theta)$$

As a general rule, the *n*th power of a complex number is derived as,

$$[r(\cos\theta + j\sin\theta)]^n = r^n(\cos n\theta + j\sin n\theta)$$

Roots of Complex Numbers

To find the roots of a complex number, take the roots of the magnitude and divide the angle,

$$\sqrt{r(\cos\theta + j\sin\theta)} = \sqrt{r}(\cos\theta/2 + j\sin\theta/2)$$

You must also keep in mind that every square root has a positive and a negative square root. The *n*th roots of a complex number has *n* number of roots, the *n* roots all have absolute value $r^{1/n}$, and the *n* values of the angle θ are obtained as,

$$\theta_n = \frac{2\pi m + \theta}{n}$$

Where *m* takes on the values from 0 to *n* – 1, it is like dividing the circle into *n* equal parts, dividing the angle into *n* parts, and adding them together.

For example, the cube root of $j = \cos\frac{\pi}{2} + j\sin\frac{\pi}{2}$ where *r* = 1, is obtained as,

$$\sqrt[3]{j} = \sqrt[3]{\cos\frac{\pi}{2} + j\sin\frac{\pi}{2}}$$

$$\sqrt[3]{j} = \left(\cos\frac{\pi}{6} + j\sin\frac{\pi}{6}\right), \left(\cos\frac{5\pi}{6} + j\sin\frac{5\pi}{6}\right), \left(\cos\frac{3\pi}{2} + j\sin\frac{3\pi}{2}\right)$$

For even n, *the roots are complex conjugates.*

THE EXPONENT e AND THE POWER FUNCTIONS

One of the most useful ways of representing a complex number is the exponential notation. There will be extensive use of exponential functions in our study of signal analysis—especially, the Fourier Transform—so it is important to grasp the concept of the exponential functions.

A product is the result of multiplying two or more quantities; if the quantities are similar, the result is a power function, in which a variable is raised to a constant power, such as area and volume,

$$Area = l \times l = l^2$$
$$Volume = l \times l \times l = l^3$$

Another form of power functions is when a constant is raised to a variable. For example, how many ways can you place two numbers in different slots? You need to know how many slots are there, as in

$$y = 2^x$$

In which x is the number of slots.

For 2 slots, the number of combination is $y = 2^2 = 4$, and for 3 slots, it is $y = 2^3 = 8$. This is also called a *growth function*: you begin with a fixed quantity and it grows. A bank's interest rate is a common application.

A simpler method for representing a power function is that in which the power factor of a base number is written as a *logarithmic number* (*log*, for short). You will find the same logarithmic value for different numbering bases. For example, the number 10^6 (1,000,000) could be written as

$$10^6 = 1000000$$
$$\log_{10}(1000000) = 6$$

And the number 2^6 (64) could be written as

$$2^6 = 64$$
$$\log_2(64) = 6$$

Except for base 1, any number can be used for a logarithmic base. It is customary to omit the base while representing a power number when the base is 10. Following are the rules of arithmetic operations on logarithmic numbers:

$$\log_a 1 = 0, a^0 = 1$$
$$\log_a a = 1, a^1 = a$$
$$\log_a(M \times N) = \log_a M + \log_a N$$
$$\log_a(\frac{M}{N}) = \log_a M - \log_a N$$
$$\log_a(N^P) = P \log_a N$$

There is a special form of growth function called *exponential growth*, in which the growth is proportional to the amount present at the time.

Examples of exponential growth include well-placed money in the stock market, as well as rate of decay of such things as radioactive materials. Rate of decay is proportional to the quantity present at the time. Things grow and the rate of growth is proportional to the quantity present at that time, such as the build up of electric charge on a capacitor, voltage buildup on an inductor, and population growth of living organisms.

Mathematically speaking, an exponential function is a power function with the rate of change proportional to the amount present at the time. In other words, with an appropriate proportionality constant, the exponent power function's derivative is equal to the power function itself. See Equation 2.3.

$$\frac{dy}{dt} \propto y,$$

$$\frac{dy}{dt} = ky \tag{2.3}$$

Let's find out what number, when raised to a power, gives us a derivative that is the same as the number itself, as in Equation 2.4:

$$\frac{d(N^t)}{dt} = N^t \tag{2.4}$$

We can determine the number experimentally by examining a few power function graphs. Let's pick three numbers—say, 2.5, 2.718, and 3—as the number base for the power functions and then calculate the values of their derivatives at different points. Table 2.1 shows the tabulation of the power functions of the three numbers

TABLE 2.1 Tabulation of the Power Functions and Their Derivatives

$t=$	$(2.5)^t$	$\dfrac{d(2.5)^t}{dt}$	$(2.718)^t$	$\dfrac{d(2.718)^t}{dt}$	$(3)^t$	$\dfrac{d(3)^t}{dt}$
3	15.63	14.32	20.08	20.08	27	29.66
3.1	17.12	15.69	22.19	22.19	30.14	33.11
3.2	18.77	17.2	24.52	24.52	33.63	36.95
3.3	20.57	18.85	27.1	27.1	37.54	41.24
3.4	22.54	20.66	29.95	29.95	41.9	46.03
3.5	24.71	22.64	33.1	33.1	46.77	51.38
3.6	27.08	24.81	36.58	36.58	52.2	57.34
3.7	29.67	27.19	40.43	40.43	58.26	64
3.8	32.52	29.8	44.68	44.68	65.02	71.43
3.9	35.64	32.66	49.38	49.38	72.57	79.73
4	39.06	35.79	54.58	54.58	81	88.99

and their derivatives, and Figure 2.6 shows how each function differs from its derivative. Comparing the graph of the derivative of the function and the function itself, we find only the number 2.718 has the slope line (dashed line) identical to the power function line (solid line). No other number function has such a property, where the derivative is the same as the function itself.

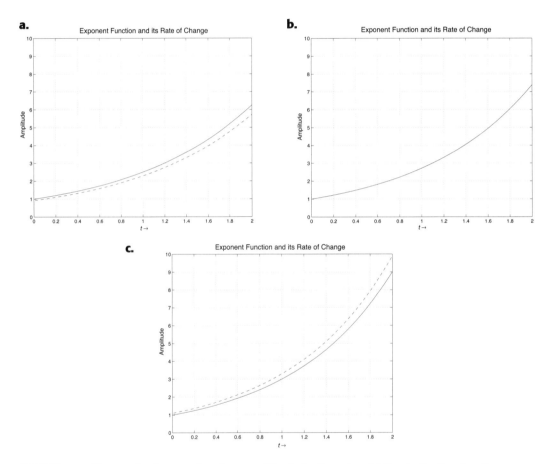

FIGURE 2.6 The graph of power functions and its derivative: a) solid line indicating function $(2.5)^t$ and dotted line indicating function $\frac{d(2.5)^t}{dt}$. b) Solid line indicating function $(2.718)^t$ and dotted line indicating function $\frac{d(2.718)^t}{dt}$. c) Solid line indicating function $(3)^t$ and dotted line indicating function $\frac{d(3)^t}{dt}$.

The number 2.718, also called the *natural number,* plays an important role in solving problems in nature involving the rate of change of quantities that follow the natural growth, as mentioned earlier. The natural number, to be more precise

2.718281828459..., is also written as e, and a function that follows exponential growth is defined as e^t. Equation 2.4 is rewritten using e notation substituted for N

$$\frac{d(e^t)}{dt} = e^t$$

A *natural logarithm* (abbreviated with ln) is a number when e is being used as a base for its logarithmic value. The natural log of e is 1, since 2.718 must be raised to the power of 1 to get 2.718, as in

$$e^1 = 2.718$$
$$\ln(e) = \ln(2.718) = 1$$

Any number raised to the power 0 is equal to 1, thus the exponent of 0 is 1 and the natural log of 1 is 0.

$$e^0 = 1$$
$$\ln(1) = 0$$

There was a remarkable discovery made by Euler (1707–1783) that established a relationship between trigonometric functions of sin and cos and the exponent function e^{jt}. This discovery greatly simplified arithmetic operation and enabled us to describe a complex number in exponent form.

Euler's Identity

Euler made an observation that the series expansion of exponent functions was equal to the series expansion of the sine and cosine function combined, as shown.

$$\cos\vartheta = 1 - \frac{\vartheta^2}{2!} + \frac{\vartheta^4}{4!} - \frac{\vartheta^6}{6!} +$$

$$j\sin\vartheta = j\vartheta - j\frac{\vartheta^3}{3!} + j\frac{\vartheta^5}{5!} - j\frac{\vartheta^7}{7!} +$$

$$\cos\vartheta + j\sin\vartheta = 1 + j\vartheta - \frac{\vartheta^2}{2!} - j\frac{\vartheta^3}{3!} + \frac{\vartheta^4}{4!} + j\frac{\vartheta^5}{5!} + ...$$

$$e^{j\vartheta} = 1 + j\vartheta - \frac{\vartheta^2}{2!} - j\frac{\vartheta^3}{3!} + \frac{\vartheta^4}{4!} + j\frac{\vartheta^5}{5!} + ...$$

We will discuss the exponent function e in depth later in the section but for now, we are interested in representing a complex number in exponent notation. Using

Euler's identity, we can denote sine and cosine functions in terms of its equivalent exponent form,

$$e^{j\vartheta} = \cos\vartheta + j\sin\vartheta$$

$$e^{-j\vartheta} = \cos\vartheta - j\sin\vartheta$$

$$\cos\vartheta = \frac{e^{j\vartheta} + e^{-j\vartheta}}{2}$$

$$\sin\vartheta = \frac{e^{j\vartheta} - e^{-j\vartheta}}{j2}$$

Going back to our complex value representation, a vector M of fixed radial line, as shown in the Figure 2.5, can be represented in the exponent form that would allow us to write a sinusoidal function using exponent notations,

$$W = Me^{j\vartheta} = M\angle^{\vartheta} = (M\cos(\vartheta) + jM\sin(\vartheta))$$

Exponent form is an ideal way of representing complex numbers when algebraic equations are involved. Following are some interesting equalities between real numbers and complex numbers, and we will use them when we derive Butterworth filters (discussed in Chapter 6, "Filter Design").

$$e^{j\pi} = \cos(\pi) + j\sin(\pi) = j^2 = -1$$

$$\left(e^{j\pi}\right)^j = \left(j^2\right)^j = \left(j^j\right)^2 = e^{-\pi} = 1/e^{\pi}$$

$$e^{j\frac{\pi}{2}} = \cos(\frac{\pi}{2}) + j\sin(\frac{\pi}{2}) = j$$

The Complex Frequencies

When a complex number $s = \sigma + j\omega$ appears in an exponential time function e^{st}, s is called the complex frequency.

$$e^{st} = e^{(\sigma + j\omega)t} = e^{\sigma t}e^{j\omega t} = e^{\sigma t}(\cos\omega t + j\sin\omega t)$$

The complex frequencies include an exponential decay factor $\sigma < 0$ (see Figure 2.7a) or exponential rise factor $\sigma > 0$ (see Figure 2.7c) multiplied with a sinusoidal function ($\cos\omega t + j\sin\omega t$). If $\sigma = 0$, the complex frequencies are pure sinusoid with no decay or rise, as shown in Figure 2.7b.

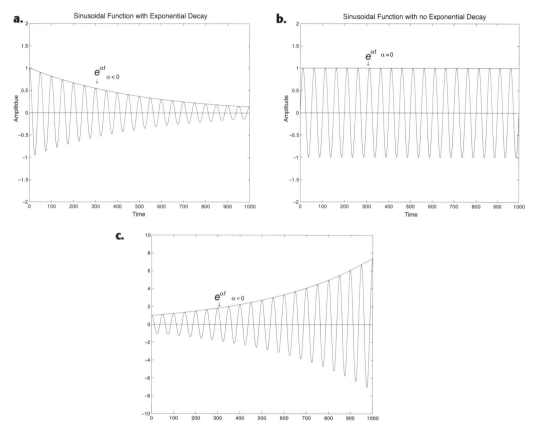

FIGURE 2.7 The varying combinations of complex sinusoids. a) Exponential decay. b) No exponent. c) Exponential rise.

Phase Angle

If a sine function starts with an angle other than 0, it has a phase angle θ that represents a time delay in the sinusoid. If ω is the frequency, ωt is the angle that is formed at the particular instance of time t. If the frequency function $\sin(\omega t)$ has a phase angle θ, the sine function must be represented as $\sin(\omega t + \theta)$. It can also be shown as a function of the complex point $W = Me^{j\theta}$ as if the line M rotates with an angular velocity ω, as shown in Figure 2.9b.

$$W(t) = Me^{j(\omega t + \theta)} = Me^{j(\omega t)} \times Me^{j(\theta)}$$

Thus, a sinusoidal function is a complex constant $Me^{j\vartheta}$ (amplitude M and phase angle θ) also known as *phasor*, multiplied by a function of time $Me^{j(\omega t)}$ indicating rotation with an angular velocity ω. Electrical circuit components such as capacitors and inductors do not alter the frequency of the input signals; they only affect the amplitude of the input wave or change the phase angle of the input frequency. The quantity $Me^{j(\omega t)}$ is usually omitted during network response calculations, since frequencies remain the same.

In terms of frequency $f = \omega/2\pi, \omega = 2\pi f$

$$W(t) = Me^{j(2\pi f + \vartheta)}$$

The projection of the line M on the real axis is

$$W_{(real)} = f(t) = (M\cos(\omega t + \vartheta))$$

And on the imaginary axis

$$W_{(imag)} = f(t) = (M\sin(\omega t + \vartheta))$$

If an instantaneous voltage is described by a sinusoidal function of time such as

$$v(t) = V\cos(\omega t + \vartheta)$$

Then $v(t)$ may be interpreted as the real part of a complex function, or

$$v(t) = \text{Re}(Ve^{j(\omega t + \vartheta)}) = \text{Re}(Ve^{j\omega t} \times Ve^{j\vartheta})$$

Figure 2.8a is a 3-dimensional space representation of a sinusoidal function showing the x-axis as the real value angular frequency ω, the y-axis as the imaginary value of $j\omega$, and time t in the z direction. If we look though the graph from the z direction, the sinusoidal function is represented as a point on a circle. A negative frequency is a function $e^{-j\omega t}$ moving in a counterclockwise direction, and a positive freqency is a function $e^{j\omega t}$ moving in a clockwise direction; both have the same amplitude and phase angle. Thus the periodic function is a function of complex numbers in continuous time,

$$y(t) = e^{j(\omega t + \vartheta)}$$
$$y(t) = e^{-j(\omega t + \vartheta)}$$

Conjugate Function

The negative exponent e^{-jt} is also called conjugate of the positive exponent e^{jt}. The two functions are mirror images of each other on a complex plane, as shown in Figure 2.8.

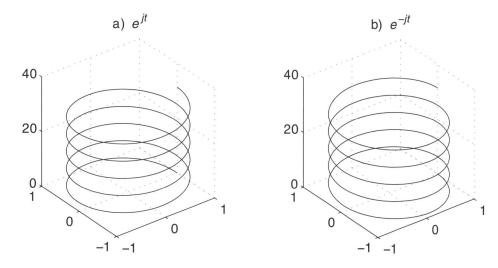

FIGURE 2.8 a) The counterclockwise direction of the function e^{jt}. b) The counterclockwise direction of the function e^{-jt}.

The advantage of phasor notation will be obvious in the next section as we try to analyze electrical networks.

THE PHASOR METHOD

The phasor method is a method of solving electrical network problems in which the current and voltage excitations applied to the networks are all sinusoidal. Before we solve these problems, a brief description of behavior of the electrical components is given as a refresher.

The Electrical Networks

Most digital signal processing was derived from the analog world, and in some situations, we would like to replace the functionality of the electrical circuits using the software algorithms, so a thorough understanding is necessary. We will study how resistors, capacitors, and inductors behave as filters of frequencies and integrate and differentiate the incoming signals of the analog world. During this study, we will assume an ideal behavior of the electrical components, where there is no loss of energy due to heat, etc.

Resistors

The voltage on a resistor is directly proportional to the current applied

$$V_r = I_r R$$

Capacitors

An electric charge builds up on the dielectric plates of a capacitor in response to an applied voltage. The charge is proportional to the voltage applied, and the capacitance C is the proportionality constant.

$$q = Cv$$

The rate of the charge buildup (known as current) is proportional to the rate of change of voltage (Equation 2.5).

$$\frac{dq}{dt} = i = C \frac{dv}{dt} \qquad (2.5)$$

For a sinusoidal excitation voltage, see Equation 2.6.

$$v = e^{j\omega t}, \frac{dv}{dt} = j\omega e^{jwt} \qquad (2.6)$$

Substituting the value of Equation 2.6 into Equation 2.5, we get

$$v = \frac{1}{j\omega C} i$$

Inductors

A rapidly changing current induces a voltage across a coil made of conductive material such as copper. The voltage is proportional to the rate of change of current, and inductance L is the proportionally constant, as shown in Equation 2.7.

$$v = L \frac{di}{dt} \qquad (2.7)$$

For a sinusoidal current, see Equation 2.8.

$$i = e^{j\omega t}, \frac{di}{dt} = j\omega e^{jwt} \qquad (2.8)$$

Substituting the value of Equation 2.8 into Equation 2.7, we get

$$v = j\omega Li$$

Table 2.2 summarizes the relationship between the input excitation to the transformed output. The impedance Z is the direct proportionality constant and the admittance Y is the inverse proportionality constant.

TABLE 2.2 Voltage and Current Relationship in Terms of Impedance and Admittance of the Three Circuit Elements

	Input excitation $v = V_m\cos(\omega t + \vartheta)$ and $i = I_m\cos(\omega t + \alpha)$	
	Impedance	**Admittance**
Resistor	$V_r = I_r R$	$I_r = V_r G$
	$R = \dfrac{1}{G}$	$G = \dfrac{1}{R}$
Capacitor	$V_c = \dfrac{1}{j\omega C} I_c$	$I_c = j\omega C V_c$
	$jX = \dfrac{1}{j\omega C}$	$jB = j\omega C$
Inductor	$V_l = j\omega L I_l$	$V_l = \dfrac{1}{j\omega L} I_l$
	$jX = \dfrac{1}{j\omega L}$	$jB = j\omega L$
	$V_l = ZI = (R + jX)I_l$	$V_l = YI = (R + jB)I_l$

EXAMPLE 2.1

Describe the voltage $v(t) = V_m \cos(\omega t + \vartheta)$ in terms of phasor.

$$v(t) = V_{in} \cos(\omega t + \vartheta)$$
$$v(t) = \text{Re}\{V_{in} e^{j(\omega t + \vartheta)}\} = \text{Re}\{(V_{in} e^{j(\vartheta)})(V_{in} e^{j(\omega t)})\}$$
$$v(t) = V_{in} e^{j(\vartheta)} = V_{in} \angle \vartheta$$

EXAMPLE 2.2

Represent the voltage $v(t) = 10\sin(\omega t\ ^\pi/_6)$ in phasor notation.

By definition, the phasors are represented as the cosine part (the real part) of the complex function in time. Since $\cos(\vartheta - ^\pi/_2) = \sin(\vartheta)$, we need to convert our sine function into cosine equivalent.

$$v(t) = 10\cos(\omega t + ^\pi/_6 - ^\pi/_2) = 10\cos(\omega t - ^\pi/_3)$$

Figure 2.9 is the graphical representation of the voltage phasors

$$V = 10 e^{-j(^\pi/_3)}$$

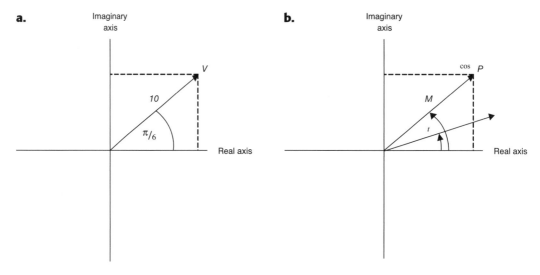

FIGURE 2.9 a) The voltage phasor of Example 2.2. b) Addition of phase angle to function $\sin(\omega t)$ as $\sin(\omega t + \theta)$.

EXAMPLE 2.3

If the current across the capacitor in a network is a function of time, as shown in Figure 2.10, write the equation of the current.

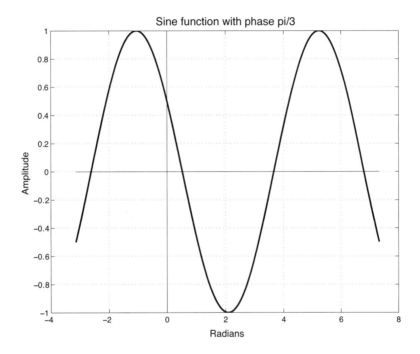

FIGURE 2.10 The voltage function for Example 2.3.

A complete cycle is achieved in 1 second, thus,

$$f = 1 cycle/\sec = \omega/2\pi$$
$$2\pi radians = 1 cycle$$
$$\omega = 2\pi radians/\sec$$

The current reaches its maximum value of 10 at 30° before the time $t = 0$, thus the phase angle

$$\vartheta = \pi/6 = 30 \deg$$

The instantaneous current is

$$i(t) = 10\cos(2\pi t + \pi/6)$$
$$i(t) = \text{Re}\{10e^{j(\omega t + \vartheta)}\} = \text{Re}\{10e^{j(2\pi t)}\} = \text{Re}\{10e^{j\vartheta}\}$$

The phasor I is

$$I = \text{Re}\{10e^{j\vartheta}\}$$
$$I = 10e^{j(\frac{\pi}{6})} \text{ Amp}$$

EXAMPLE 2.4

The sinusoidal 60 Hz AC input voltage to the circuit of Figure 2.11 is 110 V at its peak when $t = 0$. Describe the instantaneous voltage in complex notation form.

$$V_{max} = 110$$
$$\omega = 2\pi 60$$
$$\vartheta = 0$$
$$v(t) = V_{max}\cos(wt + \vartheta) = 110\cos(2\pi 60 t)$$
$$v(t) = \text{Re}\{Ve^{j(\omega t + \vartheta)}\} = \text{Re}\{(Ve^{j(\vartheta)})(e^{j(\omega t)})\}$$
$$v(t) = 110e^{j(2\pi 60 t)}$$
$$v(t) = 110\angle 0$$

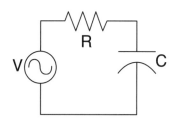

FIGURE 2.11 The voltage across the RC circuit for Example 2.4.

EXAMPLE 2.5

Find the voltage V_{out} across the capacitor C of Figure 2.11 as a function of time for an input excitation of $v(t) = V_{in}\cos(\omega t + \vartheta)$.

NOTE

Determine the current through the loop and then calculate the voltage drop across the capacitor.

The sum of the voltage drop across the resistor and the capacitor is equal to the input voltage applied.

$$V_{in} = V_R + V_C$$

$$V_{in}\cos(\omega t + \vartheta) = IR + I\frac{1}{j\omega C}$$

The current I through the loop

$$I = \frac{1}{R + \frac{1}{j\omega C}} V_{in}\cos(\omega t + \vartheta)$$

The output voltage at the capacitor

$$V_{out} = \frac{1}{j\omega C} \times \frac{1}{R + \frac{1}{j\omega C}} V_{in}\cos(\omega t + \vartheta)$$

$$V_{out} = \frac{1}{j\omega C + 1} V_{in}\cos(\omega t + \vartheta)$$

The magnitude of the output voltage is shown in Equation 2.9.

$$V_{out} = \frac{1}{\sqrt{1+(\omega RC)^2}} e^{j\vartheta} = \frac{1}{\sqrt{1+(\omega RC)^2}} \angle \tan^{-1}(\omega R \quad \quad (2.9)$$

The phase angle is shown in Equation 2.10.

$$\vartheta = \tan^{-1}(\omega RC) \quad \quad (2.10)$$

EXAMPLE 2.6

In the circuit of Figure 2.11, input voltage is a 60 Hz AC, $C = 2$ and $R = 2$. Calculate the phase angle and the voltage on capacitor C.
Substitute the values into Equation 2.9.

$$\omega = 2\pi f = 120\pi$$

$$V_c = \frac{1}{\sqrt{1 + (2 \times 2 \times 120\pi)^2}} \cos(2\pi 60 t)$$

The phase angle ϑ

$$\vartheta = \tan^{-1} \omega RC = \tan^{-1}(4 \times 120\pi)$$

EXAMPLE 2.7

Find out the frequency at which the output voltage reaches 70% of the input voltage in the circuit of Figure 2.11.
Substitute the values into Equation 2.8:

$$\frac{V_{out}}{V_{in}} = 0.7 = \frac{1}{\sqrt{1 + (j\omega RC)^2}}$$

$$0.5(1 + (\omega RC)^2) = 1$$

$$\omega RC = 1$$

The cutoff frequency is shown in Equation 2.11.

$$\omega = \frac{1}{RC} \tag{2.11}$$

NOTE
A cutoff frequency is the input frequency at which the output is 0.707 of the input.

EXAMPLE 2.8

Determine the phase angle at the cutoff frequency in the circuit of Figure 2.11.
Substituting the values into Equation 2.9:

$$\vartheta = \tan^{-1}(-1) = -45^0$$

At cutoff frequency the phase angle is 450.

Low-Pass Filter

Equation 2.9 reveals a peculiar property of the circuit in Figure 2.11: at higher frequencies, the output voltage gain is considerably lower then at lower frequencies due to the multiplying factor $\frac{1}{\sqrt{1+(j\omega RC)^2}} \times V_{in}$. The circuit could be used to discriminate lower frequencies, as if lower frequencies are passing without much degradation. The circuit of Figure 2.1 is a simple low-pass filter.

Differentiation

The circuit of Figure 2.8 is also a differentiator. The current across a capacitor is proportional to the rate of change of voltage. The same current is being fed to the resistor. Thus the output is approximately proportional to the derivative of the input voltage.

$$v = C\frac{dv}{dt}$$

EXAMPLE 2.9

A 110 Volt 60 Hz AC input is applied to the circuit of Figure 2.11, with $C = 2$ Farads and $R = 2$ Ohms. Determine the ratio of the output voltage across the capacitor to the input voltage.
Substituting the values into Equation 2.8:

$$V_{in} = 110\cos(2\pi 60 t + 0)$$

$$\frac{V_{out}}{V_{in}} = \frac{1}{\sqrt{1+(4\times 240\pi)^2}}$$

EXAMPLE 2.10

If the frequency is reduced to 1 Hz in Example 2.10, determine the improvement in the output voltage across the capacitor to the input voltage.
Substituting the values into Equation 2.8:

$$V_{in} = 110\cos(2\pi t + 0)$$
$$\frac{V_{out}}{V_{in}} = \frac{1}{\sqrt{1+(4\times 4\pi)^2}}$$

This results in nearly a 60 times improvement in the output at 1 Hz compared to the 60 Hz input.

EXAMPLE 2.11

Determine the voltage and the phase angle across the resistor R in Figure 2.12 for an input voltage of $V_{in}\cos(\omega t + \vartheta)$.

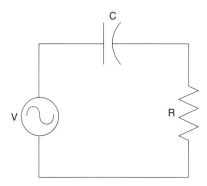

FIGURE 2.12 The voltage across the RC circuit for Example 2.11.

Similar to Example 2.5, we need to determine the current through the loop, and through that we can calculate the voltage across the resistor. The sum of the voltage drop across the resistor and the capacitor is equal to the input voltage applied.

$$V_{in}\cos(\omega t + \vartheta) = IR + I\frac{1}{j\omega C}$$

$$I = \frac{1}{R + \frac{1}{j\omega C}} V_{in}\cos(\omega t + \vartheta)$$

The magnitude of the voltage across the resistor is

$$V_{out} = \frac{1}{\sqrt{1 - \frac{1}{(\omega RC)^2}}} V_{in}\cos(\omega t + \vartheta) = \frac{1}{\sqrt{1 - \frac{1}{(\omega RC)^2}}} V_{in} \quad (2.12)$$

The phase angle is shown in Equation 2.13.

$$\vartheta = \tan^{-1}(-\frac{1}{\omega RC}) \quad (2.13)$$

High-Pass Filter

Compare Equation 2.13 with Equation 2.9. The affect is now reversed: at higher frequencies, the output voltage gain is considerably higher then at lower frequencies (the multiplying factor $\frac{1}{\sqrt{1 - \frac{1}{(\omega RC)^2}}} \times V_i$). The circuit could be used to discriminate higher frequencies, as they pass through with less degradation. The circuit of Figure 2.12 is a simple high-pass filter.

Integration

The circuit of Figure 2.7 is also an integrator, since the voltage across a capacitor gradually builds up in time. The output is approximately proportional to the integral of the input.

$$v = \frac{1}{C}\int i\,dt$$

EXAMPLE 2.12

In the circuit of Figure 2.13, inductor L is 2 mH and capacitor C is 4 uF. At what frequency will the total impedance be 0?

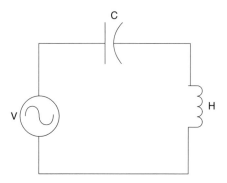

FIGURE 2.13 The *LC* circuit for Example 2.12.

The two imaginary components in the circuit are the capacitive reactance $\frac{1}{j\omega RC}$ and the inductive reactance $j\omega L$. There is no resistor, so the real component is 0.

The impedance is

$$Z = \sqrt{(0)+(\omega L - \frac{1}{\omega C})}$$

The impendence will be 0 when $\omega L = \frac{1}{\omega C}$.

$$\omega = \frac{1}{\sqrt{LC}} = \frac{1}{\sqrt{2 \times 10^{-3} \times 4 \times 10^{-6}}}$$

Resonance

Having a 0 impedance indicates an infinite gain, since the output voltage is supposed to have been divided by the impedance, which is now 0. But of course, in reality we don't have ideal components; there is a certain amount of resistance in every inductor and capacitor. But the point is, when a capacitor is connected with an inductor in a series, there is always a frequency at which both reactances just cancel each other and provide maximum gain at the output. The frequency at which the capacitive reactance is equal to the inductive reactance is called the *resonance* frequency. At resonance, the only impedance is due to the resistive components.

The resonance frequency is computed as

$$\omega = \frac{1}{\sqrt{LC}}$$

EXAMPLE 2.13

Use the phasor method to determine the current in the elements of the circuit of Figure 2.14. $C = 2F$, $R = 2\Omega$, $L = 2H$. The input voltage is $110\angle°V$ and the frequency is 1 rad/sec.

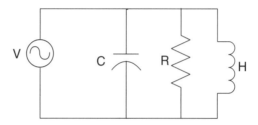

FIGURE 2.14 The *RLC* circuit for Example 2.13.

$$\omega = 1$$

$$I_R = Y_R V = \frac{1}{2} \times 110\angle^0$$

$$I_c = Y_c V = j\omega C V = j2 \times 110\angle^0 = 220\angle^{90}$$

$$I_c = Y_c V = \frac{1}{j\omega L} V = -j \times 55\angle^0 = 55\angle^{-90}$$

Notice the multiplication of *J* is the same as a phase angle of 90° and the multiplication of –*J* is the same as a phase angle of –90°.

EXAMPLE 2.14

Use the phasor method to determine the total current through the elements of the circuit of Figure 2.11 ($C = 2F, R = 2\Omega, L = 2H$). The input voltage is $110\angle°V$ and the frequency is 1 rad/sec.

The total current is the sum of the individual currents. Figure 2.15 shows the graphical addition of the current phasors.

$$I_T = I_R + I_C + I_L$$
$$I_T = I_R + I_R + I_R$$
$$I_T = 55\angle^0 + 220\angle^{90} + 55\angle^{-90}$$
$$I_T = 55 + j0 + 0 + j220 + 0 - j55$$
$$I_T = 55 + j165 = \sqrt{(55)^2 + (165)^2} \angle \tan^{-1}\frac{165}{55}$$

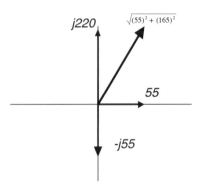

FIGURE 2.15 The graphical method of computing current phasors.

SIGNAL MODULATION

One of the most important trigonometric identities is shown in Equation 2.14.

$$\sin\theta_1 \times \sin\theta_2 = \frac{1}{2}\left[\cos(\theta_1 - \theta_2) - \cos(\theta_1 + \theta_2)\right] \quad (2.14)$$

Equation 2.14 shows that when two frequencies are multiplied, in the result we get two new frequencies: one is the sum and the other is the difference of the original frequencies. Hidden in the equation lies the essentials of communication equipment such as the radio transmitters and receivers. A radio station cannot transmit voice frequencies directly, as it would require miles of antenna. Instead for voice, a low-frequency sin wave (ranging between 100 Hz to 20000 Hz) is multiplied by a high-frequency sin wave such as 101.2 mHz(the frequency assigned to one of the radio station in the author's city), resulting in a high-frequency signal that modulates between $101.2mHz \pm 20kHz$ and $101.2mHz \pm 100Hz$. This is a frequency modulation, or *FM signal*, as shown in Figure 2.16.

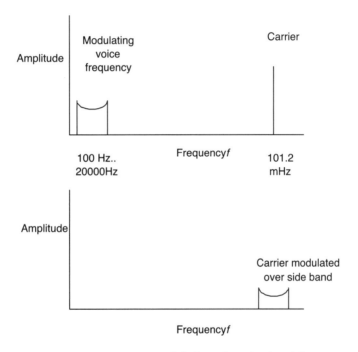

FIGURE 2.16 Frequency modulation of a voice band (100 Hz..20000Hz) with the carrier frequency (101.2mHz).

The high-frequency wave is called the *carrier wave* and the process is called voice modulated over the carrier. The sum and differences of the frequencies in Equation 2.14 are the side bands. Each sine wave in our voice creates its own side band in the modulated signal, as shown in Figure 2.16.

Frequency is not the only characteristics of the sine wave that can be modified in such a way; we could easily change the amplitude and call it *amplitude modulation* or an AM signal. Also, the phase angle could be used as the variable to modulate, which is called *phase modulation*.

Heterodyne

Transmitting the modulated high-frequency wave is easy, but on the radio receiver side, it is difficult to have a filter that can be applied equally to all ranges of high frequencies enabling you to select a particular station.

Instead, the radio receivers are provided with their own frequency multiplier that shifts the carrier frequency with their side bands to an intermediate frequency, which is away from all other radio station frequencies. An adjustable local oscillator accomplishes this task. The local oscillator is adjusted to a frequency equal to the sum and differences of the carrier and desired intermediate frequency and applied to the input signal. The process of selecting a frequency and shifting it to a new frequency is called *heterodyning*. In essence, each frequency present at the input is shifted up and down to new intermediate frequencies, from which the voice sidebands are picked up and impeded upon the speaker.

Radio transmission and reception are not the only uses of Equation 2.14. *Frequency doublers* could also be designed by multiplying a frequency to itself, as shown in Equation 2.15.

$$\sin(2\pi ft) \times \sin(2\pi ft) = \frac{1}{2} - \cos 2\pi(2 ft) \qquad (2.15)$$

The output has a dc offset, which is half the amplitude of the original wave. The dc offset could be removed by passing the signal through a capacitor acting as a high-pass filter.

The frequency doublers themselves could be used as the *phase angle detectors*. If the two frequencies are the same but differ in their phase, Equation 2.15 is modified as shown in Equation 2.16.

$$\sin(\theta) \times \sin(\theta + \vartheta) = \frac{1}{2}\cos(\vartheta) - \frac{1}{2}\cos(2\theta + \vartheta) \qquad (2.16)$$

The dc offset is now proportional to the phase difference of the input frequency as shown by the quantity $\frac{1}{2}\cos(\vartheta)$ of Equation 2.16.

We conclude the discussion of complex number arithmetic with the following summary.

SUMMARY

A complex number is a vector on a coordinate system representing amplitude and phase angle of a sinusoidal function. The two parts of a complex number are the real part (the x distance from the origin) and the imaginary part (the y distance from the origin). The imaginary part carries the operator $\sqrt{-1}$, which performs the necessary algebraic manipulations of the imaginary part. The three different ways of representing a complex number—namely the trigonometric form $(x + jy)$, the polar form $P = M\angle\vartheta$, and the exponent form $Me^{j\vartheta}$—were discussed. The phasor method was developed (based on the complex number notation) to solve the electrical network problem in which the input excitation is a sinusoidal form. The chapter provides a brief overview of the problems that our digital signal processing is supposed to solve.

3 The Fourier Transform

In This Chapter

- A Periodic Function as a Complex Number Function
- The Fourier Series of a Periodic Pulse Train
- Fourier Transforms
- Fast Fourier Transforms
- Random Signals

We discussed in Chapter 1, "Fourier Analysis," the basics of Fourier analysis and established a mathematical way of describing an arbitrary function by using its frequency components. The message was that a periodic function could be described as a composition of simple cyclic frequencies in terms of sine/cosine functions, the fundamental, and its integral multiples called harmonics. We have found complex numbers as the best way to represent points on a Cartesian coordinate system and developed the rules to perform mathematical operations using the Gaussian operator $\sqrt{-1}$. We found that a sinusoidal function can be described as a complex exponent function by using Euler's identity.

In this chapter, we will discuss the transformation of a function from time domain to frequency domain: the culmination of a Fourier series into the Fourier Transform. The frequency domain gives us another dimension of analyzing events that are extremely difficult in time domain or sometimes not even possible. We presented an example in Chapter 1 in which a mysterious output was observed when the input was a periodic function, but with the help of Fourier analysis, the problem was quickly identified. We did not include time in our analysis, and you will see that time is irrelevant when we know that the basic periodic function does not change its shape or form as time goes by.

The aim is to develop software algorithms to achieve the result of the theoretical formulation. But for that, we need to modify the Fourier formula in a manner that is suitable for digital computer implementation.

We begin by simplifying our long Fourier series representation of a function into its compact notation of Euler's exponent form and then deriving an expression that only retains the frequency information to give us a frequency domain equivalent of a time domain function. The mathematical derivative is presented to give a perspective of equivalency and show some facts of the time domain properties that are hidden in the frequency domain.

A PERIODIC FUNCTION AS A COMPLEX NUMBER FUNCTION

We have already seen that the complex exponential function as determined by Euler is a composition of sine and cosine functions.

$$e^{j\vartheta} = \cos\vartheta + j\sin\vartheta$$
$$e^{-j\vartheta} = \cos\vartheta - j\sin\vartheta$$
$$\cos\vartheta = \frac{e^{j\vartheta} + e^{-j\vartheta}}{2}$$
$$\sin\vartheta = \frac{e^{j\vartheta} - e^{-j\vartheta}}{2j}$$

A periodic function $f(t)$ in continuous time is a function of arbitrary amplitude but has a well-defined period. You can think of any function as periodic as long as you assume that the period is infinite. If you believe a function is not periodic, it may only mean that you have not waited long enough to find its period. We have seen how to extract the component frequencies in the previous chapter on Fourier analysis: the long and arduous summation of sine and cosine functions and the dc constant. Next, we'll see how to reduce this long series into a compact exponential notation using Euler's identity. The continuous time a Fourier series of a periodic function with the period $\frac{-T}{2} \le t \le \frac{T}{2}$ is presented again as a refresher in Equation 3.1:

$$f(t) = a_0 + 2\sum_{n=1}^{\infty}(a_n \sin(\frac{n2\pi t}{T}) + b_n \cos(\frac{n2\pi t}{T}))$$

Where the coefficients are

$$a_n = \frac{1}{T}\int_{-\frac{T}{2}}^{\frac{T}{2}} f(t)\sin\left(\frac{2\pi nt}{T}\right)dt$$

$$b_n = \frac{1}{T}\int_{-\frac{T}{2}}^{\frac{T}{2}} f(t)\cos\left(\frac{2\pi nt}{T}\right)dt$$

$$a_0 = \frac{1}{T}\int_{-\frac{T}{2}}^{\frac{T}{2}} f(t)dt \tag{3.1}$$

Replacing the sine and cosine functions of Equation 3.1 with the equivalent exponential notation and replacing $1/j$ with $-j$, we get,

$$\begin{aligned}
f(t) &= a_0 + \sum_{n=1}^{\infty}\frac{-ja_n}{2}\times\left(e^{\frac{j2\pi nt}{T}} - e^{-\frac{j2\pi nt}{T}}\right) + \sum_{n=1}^{\infty}\frac{b_n}{2}\times\left(e^{\frac{j2\pi nt}{T}} + e^{-\frac{j2\pi nt}{T}}\right) \\
&= a_0 + \sum_{n=1}^{\infty}\left(\frac{-ja_n + b_n}{2}\right)e^{\frac{j2\pi nt}{T}} + \left(\frac{ja_n + b_n}{2}\right)e^{-\frac{j2\pi nt}{T}} \\
&= a_0 + \sum_{n=1}^{\infty}(A_n)e^{\frac{j2\pi nt}{T}} + (B_n)e^{-\frac{j2\pi nt}{T}}
\end{aligned}$$

In which

$$A_n = \frac{b_n - ja_n}{2}, B_n = \frac{b_n + ja_n}{2}$$

Similar substitution could be made for the Fourier coefficients and the coefficients can be described using the exponential form, as in Equations 3.2 and 3.3.

$$A_n = \frac{1}{T} \int_{-\frac{T}{2}}^{\frac{T}{2}} f(t)\left(e^{-\frac{j2\pi nt}{T}}\right) dt \qquad (3.2)$$

$$B_n = \frac{1}{T} \int_{-\frac{T}{2}}^{\frac{T}{2}} f(t)\left(e^{\frac{j2\pi nt}{T}}\right) dt \qquad (3.3)$$

The remaining constant a_0 can also be defined in terms of exponent, as in

$$A_0 = A_0 e^{\frac{j2\pi 0 t}{T}} = \frac{1}{T} \int_0^T f(t) e^{0t}$$

We can merge the coefficient A_0 into A_n, since for $n = 0$ the two exponents are the same, and simplify the Fourier series some more, as shown in Equation 3.4.

$$f(t) = \sum_{n=0}^{\infty} (A_n) e^{\frac{j2\pi nt}{T}} + \sum_{n=1}^{\infty} (B_n) e^{-\frac{j2\pi nt}{T}} \qquad (3.4)$$

We could modify the second summation of Equation 3.4 so that both summations combined would cover the entire spectrum of $n = -\infty$ to $n = +\infty$, but for that, we need to redefine Bn for negative values of n only, as in Equation 3.5.

$$B_{-n} = \frac{1}{T} \int_{-\frac{T}{2}}^{\frac{T}{2}} f(t)\left(e^{-\frac{j2\pi nt}{T}}\right) dt \qquad (3.5)$$

$$f(t) = \sum_{n=-\infty}^{-1} (B_{-n}) e^{-\frac{j2\pi(-n)t}{T}} + \sum_{n=0}^{\infty} (A_n) e^{\frac{j2\pi nt}{T}}$$

Comparing Equations 3.5 and 3.2, we see that the two coefficients B_n and A_n are equal and we give them a common name C_n as we have successfully managed to combine the three coefficients of a_0, a_n, and b_n into a single form. We could also combine the two summations into one range of the values for $n = -\infty$ to $n = +\infty$; with this we can describe the Fourier series into the exponential form as follows.

Continuous time Fourier series:

$$f(t) = \sum_{n=-\infty}^{\infty} C_n e^{\frac{j2\pi nt}{T}}$$

Continuous time Fourier coefficients:

$$C_n = \frac{1}{T} \int_{-\frac{T}{2}}^{\frac{T}{2}} f(t) \times (e^{-\frac{j2\pi nt}{T}}) dt$$

$$n = -\infty, \ldots + \infty$$

The exponential functions in the previous equations are Euler's identity,

$$e^{\frac{2\pi nt}{T}} = \cos\left(\frac{2\pi nt}{T}\right) + j\sin\left(\frac{2\pi nt}{T}\right)$$

For a discrete function, the number of samples K in a period T is related with the sampling period $T = K\Delta t$ and the Nyquist frequency $N = \frac{1}{2\Delta t}$. Thus, the function $f(t)$ in time t is related to the function of discrete sampling $f(k)$ with the kth sample time $t = k\Delta t$.

We can also rewrite the discrete time Fourier series of cyclic frequency form from Chapter 1, Equation 1.6,

$$f(k) = a_0 + \sum_{n=1}^{N-1} a_n \times \sin(n2\pi k/K) + \sum_{n=1}^{N-1} b_n \times \cos(n2\pi k/K)$$

Combining the three coefficients of a_0, a_n, and b_n into a single form c_n similar to the continuous time form, we can describe the discrete time Fourier series into the compact exponential form as

$$x[k] = \sum_{n=0}^{N-1} c_n e^{\frac{j2\pi nk}{K}}$$

Similarly, the discrete version of the Fourier coefficients in the cyclic frequency form being derived in Equations 1.7, 1.8, and 1.9 of Chapter 1 can be rewritten in the compact exponential form as,

$$c_n = \frac{1}{K}\sum_{k=0}^{K-1} x[k] \times (e^{-\frac{jn2\pi k}{K}})$$

$$n = 0..K/2$$

Conventions and Notations

The normal convention of representing a function as *f(t)* is confusing with the notation for frequency as *f* so we will be using $x(t)$ to represent a continuous time, $x[k]$ to represent the discrete time function, c_n to represent discrete time coefficients, and C_n to represent continuous time Fourier coefficients. With these conventions, we rewrite the Fourier coefficient and the Fourier series formulae as follows.

Continuous time Fourier series is shown in Equation 3.6.

$$x(t) = \sum_{n=-\infty}^{\infty} C_n e^{\frac{j2\pi nt}{T}} \tag{3.6}$$

Continuous time Fourier coefficients are shown in Equation 3.7.

$$C_n = \frac{1}{T}\int_{-\frac{T}{2}}^{\frac{T}{2}} x(t) \times (e^{-\frac{j2\pi nt}{T}}) dt$$

$$n = -\infty,...+\infty \tag{3.7}$$

Discrete time Fourier series are shown in Equation 3.8.

$$x[k] = \sum_{n=0}^{N-1} c_n e^{\frac{j2\pi nk}{K}} \tag{3.8}$$

Discrete time Fourier coefficients are shown in Equation 3.9.

$$c_n = \frac{1}{K}\sum_{k=0}^{K-1} x[k] \times (e^{-\frac{jn2\pi k}{K}})$$

$$n = 0..K/2 \tag{3.9}$$

You may say that now there is a certain amount of elegance in the representation of the Fourier series and the Fourier coefficients in Equations 3.6 through 3.9, that we have taken a long series and reduced it to a compact looking function. But we certainly could not use this formula for actual calculations. We need the coefficients a_n and b_n and for that, we still have to go through the same computational process that we did with the long series. Hidden in the coefficient c_n are a_0, a_n, and b_n, and $e^{jn\omega t}$ hides the $\cos(n\omega t) + \sin(n\omega t)$ functions. Still, Equation 3.11 expresses the relationship between an input and output.

THE FOURIER SERIES OF A PERIODIC PULSE TRAIN

At this point, we would like to describe the transformation of a rectangular periodic function with unit height (as shown in Figure 3.1a) into its Fourier series. You will see the importance of this function when we discuss the FIR type filter design methods.

The pulse train as the function is being called has nonzero values only for the interval $\frac{-\Delta T}{2} \leq t \leq \frac{\Delta T}{2}$. Integrating with this limit, we obtain the coefficients of the Fourier series equivalent of the pulse train function as,

$$C_n = \frac{1}{T} \int_{-\frac{\Delta T}{2}}^{\frac{\Delta T}{2}} 1 \times e^{\frac{-j2\pi nt}{T}} dt$$

$$C_n = \frac{1}{n\pi} \sin\left(\frac{n\pi \Delta T}{T}\right)$$

The function could be rewritten as $C_n = \frac{\Delta T}{T} \times \frac{\sin\left(\frac{n\pi \Delta T}{T}\right)}{\frac{n\pi \Delta T}{T}}$ so we could use L'Hopital's rule to find the coefficient value for $n = 0$. For a very small angle, $\frac{\sin(\theta)}{\theta} = 1$. The function is of the form $\frac{\sin(\pi x)}{\pi x}$ known as a sinc(x) function. The Matlab and Scilab programs provide built-in support for sinc(x) functions. The plots in Figure 3.1b are for the number of coefficients 11, 21, and 31 for the periods $T = 1$ and $\Delta T = .2$. The listing of the Matlab program to generate the coefficients and plot the amplitude is provided on the CD-ROM.

ON THE CD

The x-axis in the plot is the coefficient number, the fundamentals, and its integer multiple harmonics. You can see the amplitude decreases as we find higher harmonics, and we discover more harmonics as the number of coefficients are increased. Now we reverse the process and bring back the original function using Equation 3.11, meaning we add all the sine and cosine functions whose amplitudes we have just calculated.

According to Equation 3.8, there are infinite harmonics in any given function, but we stopped computing after only 11 harmonics in the first case, 21 harmonics in the second case, and 31 harmonics in the third case, and the result is plotted in Figure 3.1c. The desired function is the rectangular one, and all others are approximations by using the Fourier series. As expected, the more harmonics we add, the closer we get to the desired function, but it might require an infinite number of coefficients to match the original function. The ripple effect is due to the missing sin and cosine functions that were left out of the summation. You see an extra bump right where the transition is. The magnitude is 9% higher at this point; this is known as Gibb's phenomenon.

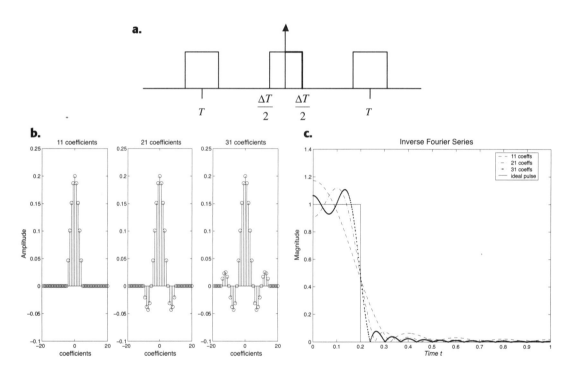

FIGURE 3.1 a) Periodic pulse train function of unit height, pulse duration ΔT, and period T. b) Fourier coefficients of the periodic pulse train function. c) Reconstruction of the function with the Fourier coefficients.

Removing the Periodic Dependency

One drawback in all Fourier analysis is that it requires the function to be periodic. The formula requires you to have data for at least one complete cycle and expects the behavior to be repeated. But it is often difficult to determine what a wave looked like earlier and what it will look like later—or even if it is really a periodic wave at all. After all, nature is too complex to have a repeated pattern. This is when the Fourier Transform is applied, which does not depend on the function being periodic. The Fourier Transform is based upon the following principle: a sample from a lot tells you something about a lot, and more samples from the lot tell you more about the lot, so let's modify the Euler's formula and remove the periodic dependency and see the consequences of it on the original function.

Euler's Equation and Nonperiodic Waves

The Fourier series in its original form is not very practical. It requires tremendous amounts of calculations, and it assumes that the input function is periodic in nature. Euler's formula—or the complex number representation of the Fourier series, shown in Equation 3.8—has hidden the conditionality of the wave being periodic in its coefficient c_n, but leaves us with a dilemma. We cannot bring back the original function, as it does not give us the actual coefficients a_0, a_n, and b_n that we need. In this section, we will modify the Euler's formula so we can deal with any kind of wave, periodic or nonperiodic.

To begin our discussion, let's assume all functions are periodic in nature. We assume every function has a period and its value is infinite. According to the Fourier series, to get a complete spectrum of the frequencies in a function, we need to observe the function for one complete cycle. Does it mean the job can never be completed? A period that approaches infinity raises other problems, too. It would make the fundamental harmonics near 0 as shown in the Equation 3.10.

$$\Delta f = \frac{1}{T_{\to\infty}} \to \frac{1}{\infty} \to 0 \qquad (3.10)$$

The more samples you take in a one-second duration, the more you expand your frequency spectrum. Every sample contributes to some degree towards the whole—even a single sample—and brings you closer in defining the frequency components of the original function whose period is now assumed to be infinite.

FOURIER TRANSFORMS

The argument presented in the previous section leads us to conclude that the factor $1/T$ must be removed from our calculations for the Euler's formula to have any practical value. If the wave takes infinite time to advance through one period, its frequency approaches 0. What happens to the spectrum of the frequency, as the wave passes through only the smallest fraction of its period in one second? The spectrum shows only the integral multiple of the fundamental frequency, so as the frequency approaches 0, the interval between frequencies on the spectrum narrows. Eventually as the period approaches infinity, the spectrum becomes a continuum and all frequencies are accounted for and there is no gap between the frequencies. Because every sample is considered a contributing factor towards the frequency component and the number of samples is related to the number of frequencies in our spectrum, it is no longer a time domain. The domain is our frequency spectrum and the rate of sampling is Δf instead of Δt.

$$\Delta f = \frac{1}{T}$$

The graph in Figures 3.2a, b, and c shows how rate of sampling affects the frequency spectrum. The gap between frequency Δf approaches 0 as the number of samples in one second approaches infinity.

Having realized that frequency is a continuum function, the fundamental and the harmonics nf_0 are simply a continuous frequency variable f with no gaps among them. Let's rewrite Equation 3.7 by replacing $\frac{1}{T}$ with df (as $T \to \infty$) and $\frac{n2\pi t}{T}$ with $2\pi f$, C_n with $X(f)$; see Equation 3.11.

$$X(f) = df \times \int_{n=-\frac{T}{2}}^{\frac{T}{2}} x(t) \times e^{-j2\pi ft} \times dt \tag{3.11}$$

Where $x(t)$ is from Equation 3.6, we get Equation 3.12.

$$x(t) = \sum_{n=-\infty}^{\infty} X(f) \times (e^{j2\pi ft}) \tag{3.12}$$

Replacing the value of $X(f)$ in Equation 3.12 with the Equation 3.11, gives us Equation 3.13.

$$x(t) = \sum_{n=-\infty}^{\infty} \left\{ df \times \int_{n=-\frac{T}{2}}^{\frac{T}{2}} x(t) \times e^{-j2\pi ft} \times dt \right\} (e^{j2\pi ft}) \tag{3.13}$$

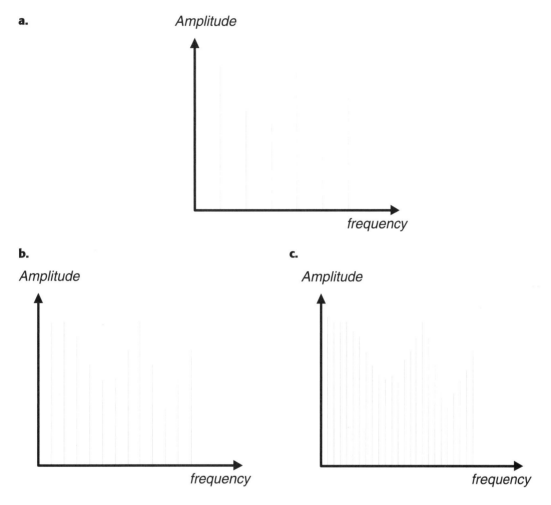

FIGURE 3.2 a) Frequency spectrum with 6 samples/sec. b) Frequency spectrum with 12 samples/sec. c) Frequency spectrum with 24 samples/sec.

The sticky point in the Equation 3.13 is the multiplying factor *df*, which is very close to 0, so we will move it out, as in Equation 3.14.

$$x(t) = \sum_{n=-\infty}^{\infty} \left\{ \int_{-\infty}^{\infty} x(t) \times e^{-j2\pi ft} \times dt \right\} (e^{j2\pi ft}) \times df \qquad (3.14)$$

We have simply taken Equation 3.14 and moved the quantity *df* to the end. We have also replaced the quantity $\frac{T}{2}$ with ∞, since T is infinite so $\frac{T}{2}$ is also infinite.

The term in the bracket of Equation 3.14 is our Fourier Transform. As we intended, it does not depend upon information regarding the period. Now is the time to change the outer summation of Equation 3.14 into an integration, since we have a very small df to integrate with (Equation 3.15).

$$x(t) = \int_{-\infty}^{\infty} \left\{ \int_{-\infty}^{\infty} x(t) \times e^{-j2\pi ft} \times dt \right\} (e^{j2\pi ft}) \times df \tag{3.15}$$

The Fourier Transform is shown in Equation 3.16.

$$X(f) = \int_{-\infty}^{\infty} x(t) \times e^{-j2\pi ft} \times dt \tag{3.16}$$

The inverse Fourier Transform is shown in Equation 3.17.

$$x(t) = \int_{-\infty}^{\infty} X(f) \times (e^{j2\pi ft}) \times df \tag{3.17}$$

Notice the removal of the term $1/T$ from Equation 3.7 of the Fourier coefficient formula, it got shifted toward the end in Equation 3.15. To bring back the original function, we can apply an inverse Fourier Transform by substituting the value $X(f)$ back in Equation 3.17.

But the Fourier Transform performed in Equation 3.16 is not the actual frequency component. The actual frequency component is our C_n. Let's compare the two.

$$C_n = \frac{1}{T} \int_{-\frac{T}{2}}^{\frac{T}{2}} x(t) \times (e^{-\frac{j2\pi nt}{T}}) dt$$

$$n = -\infty, \ldots + \infty$$

$$x(f) = \int_{-\infty}^{\infty} X(t) \times e^{-j2\pi ft} \times dt$$

The two quantities are very much alike. But C_n is the true area divided by the length as given by the term $\frac{1}{T}$, whereas $X(f)$ is a relative term. It may not show the actual frequency but it is unique for every frequency in the function. And this is good enough for our analysis.

We can take a complex wave and identify its component objects with a Fourier Transform of Equation of 3.16 (we cannot call them sine and cosine frequencies as explained earlier) and with an inverse Fourier Transform the same component objects combined return our original function as in Equation 3.17.

The Difference Between Fourier Transforms and Fourier Series

The Fourier Transform gives us a comparative analysis of component frequencies, whereas Fourier series give us the true analysis of component frequencies. The Fourier series has Fourier coefficients to get back to the original function and in Fourier Transforms the transform itself is being used to bring back the original function.

We have Fourier coefficients in an acceptable form in Equation 3.16 in the form of a Fourier Transform. The integral in Equation 3.16 is an improper integral since the domain of integration is an unbound interval. The convergence or divergence of the integral depends entirely on the function $x(t)$, since the magnitude of the term $e^{-j2\pi ft}$ never exceeds 1.

$$\left|e^{-j2\pi ft}\right| = \left|\cos(2\pi ft) - j\sin(2\pi ft)\right| = (\cos^2(2\pi ft) + \sin^2(2\pi ft)) = 1$$

The Fourier Transform of a function shows us the magnitude of the component frequencies present in a function, but how close we get in approximating the true contents is a topic of discussion next.

You will see the benefits of Fourier Transforms from a digital signal processing point of view later in the section, but a little bit more rigor in mathematics is presented next to show you a brilliant conclusion of the Fourier Transform : the Heisenberg Uncertainty Principal.

The Difference Between a True Function and the Estimation with the Fourier Transform

Let's analyze the result for a moment and see how good our estimation is with our Fourier Transform. In an ideal situation, if we have a wave that has an infinite period and find the Fourier Transform of it, we should see only one frequency $f = 0$. The function $g(t) = 1$ is one example that does not oscillate. It goes from –ve to +ve infinity without touching the ground. It has only one frequency $f = 0$. See Figure 3.3.

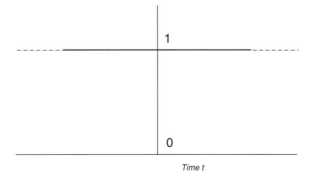

FIGURE 3.3 A function with a constant value of 1.

If we apply the Fourier Transform and analyze the spectrum, it should look like the frequency graph of Figure 3.4. That means there should be only one frequency at 0.

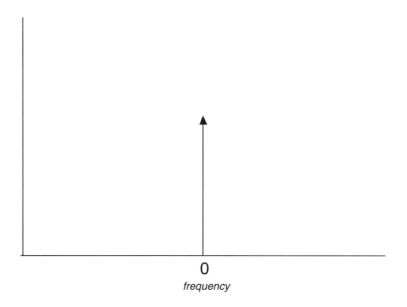

FIGURE 3.4 Frequency spectrum of an ideal function with a constant value.

It would be next to impossible to integrate this function over a period, as the period stretches all the way up to infinity, so we take a finite time period of ΔT and try to integrate the function as shown in Figure 3.5. We have to assume that beyond ΔT everywhere the function is 0. It is different from the previous example of the pulse train where the pulse is a periodic function of the period T.

Integrating Equation 3.18 for the range $\frac{-\Delta T}{2} \leq t \leq \frac{\Delta T}{2}$, we obtain the Fourier Transform of the function $g(t) = 1$ as follows,

$$X(f) = \int_{-\frac{\Delta T}{2}}^{\frac{\Delta T}{2}} 1 \times e^{-j2\pi ft} dt$$

$$X(f) = \frac{\sin(\pi f \Delta T)}{\pi f}$$

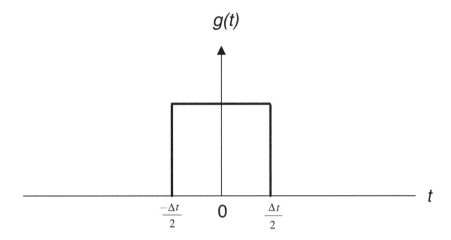

FIGURE 3.5 A finite time sampling of an infinite duration function.

The function could be rewritten as $X(f) = \Delta T \times \frac{\sin(\pi f \Delta T)}{\pi f \Delta T}$, a sinc($x$) function, similar to the Fourier coefficients. But the difference is that the sinc(x) function in the Fourier coefficient formula had a multiplier $\frac{\Delta T}{T}$, whereas the Fourier Transform has only ΔT. To judge our estimate, we pick three different ΔT values as sampling intervals to estimate our constant function $g(t) = 1$. The greater the ΔT means more time is being spent observing the event. As usual, we expect only one frequency at $f = 0$, but as in the Fourier series example, we get other frequency contents besides $f = 0$, though with diminishing effect as our sampling time gets larger. The thing that is different from the Fourier series is that the height of the graph grows with the width of the sampling period ΔT. Figure 3.6 shows the result for $\Delta T = 2$, $\Delta T = 4$, and $\Delta T = 6$.

$$g(t) = \begin{vmatrix} c\,for\,|x| \leq t \\ 0\,for\,|x| \geq -t \end{vmatrix}$$

We reach an interesting conclusion. The amplitude of the true frequency increases as we increase our period of samples. (Assuming the function is a constant with an infinite period and there is only one frequency at 0 in the function.) We should be seeing only one component in our frequency spectrum and it should have the highest value at 0 frequency, as in Figure 3.6. But the result in Figure 3.6 shows a trend toward true frequency of 0, but there are other frequencies also in our spectrum. To achieve the true result of 0 frequency, the duration of sample ΔT should have been infinite.

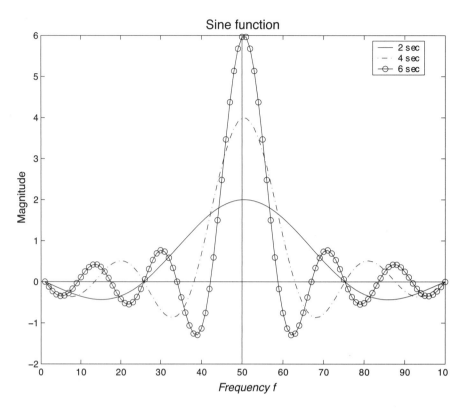

FIGURE 3.6 The estimation of the unit function $g(t) = 1$ with the Fourier Transform using sampling interval $\Delta T = 2$, $\Delta T = 4$, and $\Delta T = 6$.

Thus, the more we observe, the better we understand. Every new observation brings us closer to a new reality about the function. And to achieve perfection, we should keep observing forever but as that is not possible, we never achieve perfection. Every time we stop to consolidate our gain, we add a little unknown in the formation of the function. Just like with the Fourier Transform, there is a little bit of uncertainty in our estimation of a true function.

If ΔT is a measure of a sample in time, then Δf could be a measure of uncertainty in our sample. The Fourier Transform tells us that the quantity $\Delta T \times \Delta f$ is a constant, and to reduce the gap in frequencies (reduce the Δf), we must increase our time of sample ΔT. There is a relationship of variance—or the rate of change of sample in time—and the variance in its Fourier Transform, which is given by Werner Heisenberg as the uncertainty principle.

$$Var\{h(t)\} \times Var\{H(t)\} \geq \frac{1}{16\pi^2}$$

Discrete Time Fourier Transforms

We now turn our attention to the problem of implementing an algorithm to compute the Fourier Transform using a digital computer. The method is essentially a sampling of events taken at a discrete time. The mathematics had to be modified slightly to take into account that only at a specific time does the discrete time coincide with real time. We only need two subsequent samples and the time duration between the samples to identify a particular frequency, as explained by the Nyquist criteria of Chapter 1.

Each discrete sample of continuous time function $g(t)$ at the interval Δt coincides at the kth position of the function and is represented as $x[k]$. If ω is the angular velocity in radians per second, then $\omega \Delta t$ is the advancement of the wave in radians per sample.

If K is the number of samples per second, the fundamental period is $K\Delta t$. The fundamental frequency is given as $f = \frac{1}{K\Delta t}$, and the nth frequency component is $f_n = n/\Delta t K$, where $n = 0,1,2,3,4...N$ (Nyquist frequency). Since t coincides with $k\Delta t$, the discrete version of the exponent term in the Fourier coefficient formula of Equation 3.18 may be defined as,

$$e^{-j2\pi f dt} = e^{-j2\pi k \Delta t \frac{n}{\Delta t K}}$$

Modifying the Fourier Transform of Equation 3.18 for discrete time events, we have discrete time Fourier Transform as follows,

$$X(n) = \Delta t \times \sum_{k=0}^{K-1} x[k] \times e^{-j2\pi k \frac{n}{K}}$$

Since Δt is our choice, we could normalize the equation by choosing $\Delta t = 1$ and simplify the Fourier Transform formula as Equation 3.18 shows.

The discrete Fourier Transform $X(n)$ of a discrete sampling function $x[k]$ is shown in Equation 3.18.

$$X(n) = \sum_{k=0}^{K-1} x[k] \times e^{-j2\pi k \frac{n}{K}} \tag{3.18}$$

Where the exponential term is,

$$e^{-j2\pi k \frac{n}{K}} = \cos(2\pi k \frac{n}{K}) - j\sin(2\pi k \frac{n}{K})$$

Similarly, the inverse Fourier Transform given in Equation 3.17 may be converted into its discrete version, but there is an extra term df at the end that is defined as the rate of sampling. We have already decided to normalize the sampling period with $\Delta t = 1$, then $df = \frac{1}{K\Delta t} = \frac{1}{K}$.

The inverse discrete Fourier Transform to recover the sampling function $x[k]$ from the Fourier Transform $X(n)$ may be defined as in Equation 3.19.

$$x[k] = \frac{1}{K}\sum_{k=0}^{K-1} X(n) \times e^{j2\pi k \frac{n}{K}} \tag{3.19}$$

The only two things you need to define before processing is the sampling rate Δt and the number of samples K in the processing, and Equation 3.18 will compute all the harmonics from 0 to $K/2$.

Equation 3.18 is computationally intensive as is. Just to give you a real life example of the number of computations involved, let's analyze a voice spectrum on digitized data of sound pattern.

As a first step, we need to determine the maximum frequency that needs to be extracted from the sampled data. The human voice can easily reach up to 4000 Hz. That means we need at least 8,000 data points per second. We have $K = 8000$, $N = 4000$, and $\Delta t = 1/8000$.

The number of computations is $2 \times 4000 \times 8000$ just to get the magnitude of the Fourier Transform, a very impractical suggestion. Fortunately, there are ways to improve the algorithm and that is being discussed next in "Fast Fourier Transforms."

FAST FOURIER TRANSFORMS

Reducing the number of computations in the discrete Fourier Transform has been a dream of many and if you take a closer look, you will see that there are a lot of redundant calculations. It becomes clear when we rewrite Equation 3.18 in a slightly different format, as in Equation 3.20.

$$X(n) = \sum_{k=0}^{K-1} x[k] \times e^{-j\frac{2\pi nk}{K}} \tag{3.20}$$

The exponential term $e^{-j\frac{2\pi}{K}}$ in Equation 3.20 is a complex number indicating a point on a circle. The fact that it is multiplied by nk shows the point is being rotated around the circle nk times. The $-j$ only signifies that the rotation is in clockwise direction. Let's assume that you are computing a Fourier Transform of eight samples, $K = 8$, hence the number of frequencies $n = 4$ and you step through each frequency $k = 0$ seven times. That means the complex point is rotated $nk = 32$ times around the circle. But after the first eight rotations, the point simply moves around the circle and repeats its pattern, as shown in Figure 3.7.

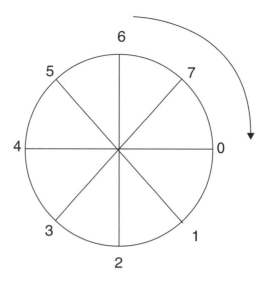

FIGURE 3.7 A point $e^{-j\frac{2\pi}{N}}$ moves around the circle when multiplied by *nk*.

For every new wave $n = 0,1,2,3$, the multiplication steps are repeated for $k = 0,1,2,3,4,5,6,7$, but notice the point never leaves the circle. In essence, $e^{-j\frac{\pi}{4}\times 0}$ is equal to $e^{-j\frac{\pi}{4}\times 8}$, and $e^{-j\frac{\pi}{4}\times 1}$ is equal to $e^{-j\frac{\pi}{4}\times 9}$, etc. Table 3.1 shows the actual complex number vector involved once redundancies are removed. The column headings are the numerical points and the row headings are the frequencies.

Look carefully at Table 3.1 and you will see that every even point has a top and bottom half that are identical. It means we only need to compute the top half and use the same computations for the bottom half. This is one big savings in multiplication operation. Now there is a different set of rules: if $k = 0,1,2,3,4,5,6,7$, then $2k$ are the even points $0,2,4,6$, and $2k + 1$ are the odd points $1,3,5,7$.

We can rewrite Equation 3.20 to reflect the new set of operations as shown in Equation 3.21.

$$X(n) = \Delta t \times \{\{\sum_{k=0}^{K/2-1} x[2k] \times e^{-j\frac{2\pi}{K}\times n(2k)}\} + \{\sum_{k=0}^{K/2-1} x[2k+1] \times e^{-j\frac{2\pi}{K}\times n(2k+1)}\}\} \quad (3.21)$$

Let's split the group into odd and even operations and rearrange the table to reflect the new set of computations, as shown in Table 3.2.

TABLE 3.1 Complex Numbers to Be Multiplied in an Eight-point DFT

	x(0)	x(1)	x(2)	x(3)	x(4)	x(5)	x(6)	x(7)
X(0/8)	$e^{-j\frac{\pi}{4}\times 0}$	$e^{-j\frac{\pi}{4}\times 0}$	$e^{-j\frac{\pi}{4}\times 0}$	$e^{-j\frac{\pi}{4}\times 0}$	$e^{-j\frac{\pi}{4}\times 0}$	$e^{-j\frac{\pi}{4}\times 0}$	$e^{-j\frac{\pi}{4}\times 0}$	$e^{-j\frac{\pi}{4}\times 0}$
X(1/8)	$e^{-j\frac{\pi}{4}\times 0}$	$e^{-j\frac{\pi}{4}\times 1}$	$e^{-j\frac{\pi}{4}\times 2}$	$e^{-j\frac{\pi}{4}\times 3}$	$e^{-j\frac{\pi}{4}\times 4}$	$e^{-j\frac{\pi}{4}\times 5}$	$e^{-j\frac{\pi}{4}\times 6}$	$e^{-j\frac{\pi}{4}\times 7}$
X(2/8)	$e^{-j\frac{\pi}{4}\times 0}$	$e^{-j\frac{\pi}{4}\times 2}$	$e^{-j\frac{\pi}{4}\times 4}$	$e^{-j\frac{\pi}{4}\times 6}$	$e^{-j\frac{\pi}{4}\times 0}$	$e^{-j\frac{\pi}{4}\times 2}$	$e^{-j\frac{\pi}{4}\times 4}$	$e^{-j\frac{\pi}{4}\times 6}$
X(3/8)	$e^{-j\frac{\pi}{4}\times 0}$	$e^{-j\frac{\pi}{4}\times 3}$	$e^{-j\frac{\pi}{4}\times 6}$	$e^{-j\frac{\pi}{4}\times 1}$	$e^{-j\frac{\pi}{4}\times 4}$	$e^{-j\frac{\pi}{4}\times 7}$	$e^{-j\frac{\pi}{4}\times 2}$	$e^{-j\frac{\pi}{4}\times 5}$
X(4/8)	$e^{-j\frac{\pi}{4}\times 0}$	$e^{-j\frac{\pi}{4}\times 4}$	$e^{-j\frac{\pi}{4}\times 0}$	$e^{-j\frac{\pi}{4}\times 4}$	$e^{-j\frac{\pi}{4}\times 0}$	$e^{-j\frac{\pi}{4}\times 4}$	$e^{-j\frac{\pi}{4}\times 0}$	$e^{-j\frac{\pi}{4}\times 4}$
X(5/8)	$e^{-j\frac{\pi}{4}\times 0}$	$e^{-j\frac{\pi}{4}\times 5}$	$e^{-j\frac{\pi}{4}\times 2}$	$e^{-j\frac{\pi}{4}\times 7}$	$e^{-j\frac{\pi}{4}\times 4}$	$e^{-j\frac{\pi}{4}\times 1}$	$e^{-j\frac{\pi}{4}\times 6}$	$e^{-j\frac{\pi}{4}\times 3}$
X(6/8)	$e^{-j\frac{\pi}{4}\times 0}$	$e^{-j\frac{\pi}{4}\times 6}$	$e^{-j\frac{\pi}{4}\times 4}$	$e^{-j\frac{\pi}{4}\times 2}$	$e^{-j\frac{\pi}{4}\times 0}$	$e^{-j\frac{\pi}{4}\times 6}$	$e^{-j\frac{\pi}{4}\times 4}$	$e^{-j\frac{\pi}{4}\times 2}$
X(7/8)	$e^{-j\frac{\pi}{4}\times 0}$	$e^{-j\frac{\pi}{4}\times 7}$	$e^{-j\frac{\pi}{4}\times 6}$	$e^{-j\frac{\pi}{4}\times 5}$	$e^{-j\frac{\pi}{4}\times 4}$	$e^{-j\frac{\pi}{4}\times 3}$	$e^{-j\frac{\pi}{4}\times 2}$	$e^{-j\frac{\pi}{4}\times 1}$

TABLE 3.2 Complex Numbers to Be Multiplied in an Eight-point DFT with Blank Entries Indicating Redundant Operations

	x(0)	x(1)	x(2)	x(3)	x(4)	x(5)	x(6)	x(7)
X(0/8)	$e^{-j\frac{\pi}{4}\times 0}$	$e^{-j\frac{\pi}{4}\times 0}$	$e^{-j\frac{\pi}{4}\times 0}$	$e^{-j\frac{\pi}{4}\times 0}$	$e^{-j\frac{\pi}{4}\times 0}$	$e^{-j\frac{\pi}{4}\times 0}$	$e^{-j\frac{\pi}{4}\times 0}$	$e^{-j\frac{\pi}{4}\times 0}$
X(1/8)	$e^{-j\frac{\pi}{4}\times 0}$	$e^{-j\frac{\pi}{4}\times 2}$	$e^{-j\frac{\pi}{4}\times 4}$	$e^{-j\frac{\pi}{4}\times 6}$	$e^{-j\frac{\pi}{4}\times 1}$	$e^{-j\frac{\pi}{4}\times 3}$	$e^{-j\frac{\pi}{4}\times 5}$	$e^{-j\frac{\pi}{4}\times 7}$
X(2/8)	$e^{-j\frac{\pi}{4}\times 0}$	$e^{-j\frac{\pi}{4}\times 4}$	$e^{-j\frac{\pi}{4}\times 0}$	$e^{-j\frac{\pi}{4}\times 4}$	$e^{-j\frac{\pi}{4}\times 2}$	$e^{-j\frac{\pi}{4}\times 6}$	$e^{-j\frac{\pi}{4}\times 2}$	$e^{-j\frac{\pi}{4}\times 6}$
X(3/8)	$e^{-j\frac{\pi}{4}\times 0}$	$e^{-j\frac{\pi}{4}\times 6}$	$e^{-j\frac{\pi}{4}\times 4}$	$e^{-j\frac{\pi}{4}\times 2}$	$e^{-j\frac{\pi}{4}\times 3}$	$e^{-j\frac{\pi}{4}\times 1}$	$e^{-j\frac{\pi}{4}\times 7}$	$e^{-j\frac{\pi}{4}\times 5}$
X(4/8)					$e^{-j\frac{\pi}{4}\times 4}$	$e^{-j\frac{\pi}{4}\times 4}$	$e^{-j\frac{\pi}{4}\times 4}$	$e^{-j\frac{\pi}{4}\times 4}$
X(5/8)					$e^{-j\frac{\pi}{4}\times 5}$	$e^{-j\frac{\pi}{4}\times 7}$	$e^{-j\frac{\pi}{4}\times 1}$	$e^{-j\frac{\pi}{4}\times 3}$
X(6/8)					$e^{-j\frac{\pi}{4}\times 6}$	$e^{-j\frac{\pi}{4}\times 2}$	$e^{-j\frac{\pi}{4}\times 6}$	$e^{-j\frac{\pi}{4}\times 2}$
X(7/8)					$e^{-j\frac{\pi}{4}\times 7}$	$e^{-j\frac{\pi}{4}\times 5}$	$e^{-j\frac{\pi}{4}\times 3}$	$e^{-j\frac{\pi}{4}\times 1}$

We have one more trick of mathematics that we can apply to Equation 3.21. The odd complex number multiplier $(2k + 1)$ can be converted into an even complex number by separating the exponent as follows,

$$e^{-j\frac{n\pi}{N}(2k+1)} = e^{-j\frac{\pi}{N}n(2k)} \times e^{-j\frac{\pi}{N}n}$$

What we have done essentially is rotated the vector that makes the point fall on to the next even location. The operation will become clear when we discuss vector rotation in the next section, but for now, just see the advantage of converting an odd complex number operation into an even complex number operation. We can apply the same divide-and-conquer rule that we did in the first part of Table 3.2 where even numbers' top and bottom halves became identical.

The new equation is

$$X(n) = \Delta t \times \{\{\sum_{k=0}^{K/2-1} x[2k] \times e^{-j\frac{2\pi}{K} \times n(2k)}\} + e^{-j\frac{2\pi}{K} \times n} \times \{\sum_{k=0}^{K/2-1} x[2k+1] \times e^{-j\frac{2\pi}{K} \times n(2k)}\}\}$$

If we expand the operation on the righthand side of the equation into a table, as we did in the original computation in Table 3.2, we will see the same repeated pattern emerging in tabulation. No entries are made in the bottom half of Table 3.3 since the bottom half is identical to the top half. But don't forget that at the end of all multiplication and summation we need to perform one last operation of multiplication with $e^{-j\frac{2\pi}{K}n}$.

$$e^{-j\frac{2\pi}{K} \times n} \times \{\sum_{k=0}^{K/2-1} x[2k+1] \times e^{-j\frac{2\pi}{K} \times n(2k)}\}\}$$

$$K = 0,1,2,3$$

$$n = 0,1,2,3,4,5,6,7$$

Let's divide and conquer some more, since we are doing so well. Look at every alternate point in Table 3.3. The top two and the bottom two are the same just like in Table 3.1. We can eliminate the redundant calculations and reduce computations some more as in Table 3.4.

TABLE 3.3 The Complex Number Multiplier in Odd Points Calculations

	x(1)	x(3)	x(5)	x(7)
X(0/8)	$e^{-j\frac{\pi}{4}\times 0}$	$e^{-j\frac{\pi}{4}\times 0}$	$e^{-j\frac{\pi}{4}\times 0}$	$e^{-j\frac{\pi}{4}\times 4}$
X(1/8)	$e^{-j\frac{\pi}{4}\times 0}$	$e^{-j\frac{\pi}{4}\times 2}$	$e^{-j\frac{\pi}{4}\times 4}$	$e^{-j\frac{\pi}{4}\times 6}$
X(2/8)	$e^{-j\frac{\pi}{4}\times 0}$	$e^{-j\frac{\pi}{4}\times 4}$	$e^{-j\frac{\pi}{4}\times 0}$	$e^{-j\frac{\pi}{4}\times 4}$
X(3/8)	$e^{-j\frac{\pi}{4}\times 0}$	$e^{-j\frac{\pi}{4}\times 6}$	$e^{-j\frac{\pi}{4}\times 4}$	$e^{-j\frac{\pi}{4}\times 2}$

TABLE 3.4 The Complex Number Multiplier in Odd Points Calculations, Blank Entries Indicating Redundant Operation

	x(1)	x(5)	x(3)	x(7)
X(0/8)	$e^{-j\frac{\pi}{4}\times 0}$	$e^{-j\frac{\pi}{4}\times 0}$	$e^{-j\frac{\pi}{4}\times 0}$	$e^{-j\frac{\pi}{4}\times 4}$
X(1/8)	$e^{-j\frac{\pi}{4}\times 0}$	$e^{-j\frac{\pi}{4}\times 4}$	$e^{-j\frac{\pi}{4}\times 2}$	$e^{-j\frac{\pi}{4}\times 6}$
X(2/8)			$e^{-j\frac{\pi}{4}\times 4}$	$e^{-j\frac{\pi}{4}\times 4}$
X(3/8)			$e^{-j\frac{\pi}{4}\times 6}$	$e^{-j\frac{\pi}{4}\times 2}$

We keep rotating and eliminating the redundant computations until we have only one point left. Let's make some simple substitutions to make the expressions a little cleaner and remove the redundancies while expanding the Fourier Transform computations as shown in Equation 3.22.

$$x[2k] = p[k]$$
$$x[2k+1] = q[k]$$

$$e^{-j\frac{\pi}{N}} = W$$

$$X(n) = \Delta t \times \{\{\sum_{k=0}^{K/2-1} p(k)\times W^{n(2k)}\} + W^n \times \{\sum_{k=0}^{K/2-1} q(k)\times W^{n(2k)}\}\}$$

$$X(n) = \Delta t \times \{\sum_{k=0}^{K/4-1} p(2k) \times W^{'n(2k)}\} + W^{'n} \times \{\sum_{k=0}^{K/4-1} p(2k+1) \times W^{'n(2k)}\}$$

$$+ W^n \{\sum_{k=0}^{K/4-1} p(2k) \times W^{'n(2k)}\} + W^{'n} \times \{\sum_{k=0}^{K/4-1} q(2k+1) \times W^{'n(2k)}\}\} \quad (3.22)$$

Let's see the total savings we achieve by rotating the point and eliminating the redundant computations,

Step 1: $a(0), a(1) \rightarrow 4$

Step 2: $W^{'n} \times (b(0), b(1)) \rightarrow 4 + 2$

Step 3: $W^n \times (c(0), c(1)) \rightarrow 4 + 2$

Step 4: $W^n[W^{'n} \times (d(0), d(1))] \rightarrow 4 + 2 + 2$

Total = 24

The original discrete Fourier Transform required $8 \times 8 = 64$ multiplications that were reduced to only 24 multiplications in the Fast Fourier Transform algorithms. It is a savings of a magnitude in time.

The multiplication operation with $e^{nk\frac{\pi}{K}}$ can be explained as an operation of a vector rotation.

The table and the algorithm were obtained from the publication Who's Fourier.

Vector Rotation

A complex number $e^{-j\alpha}$ is a vector in a Cartesian coordinate system with a magnitude R and a phase angle α. Multiplying this vector with another complex number $e^{-j\beta}$ is like rotating through an angle β as shown in Figure 3.8a.

The X component of the vector $X = R\cos(\alpha)$, and the rotation of the X component through an angle β is $X\cos(\beta)$ and $X\sin(\beta)$, as shown in Figure 3.8b.

The Y component of the vector $Y = R\sin(\alpha)$, and the rotation of the Y component through an angle β is $Y\cos(\beta)$ and $Y\sin(\beta)$, as shown in Figure 3.8c.

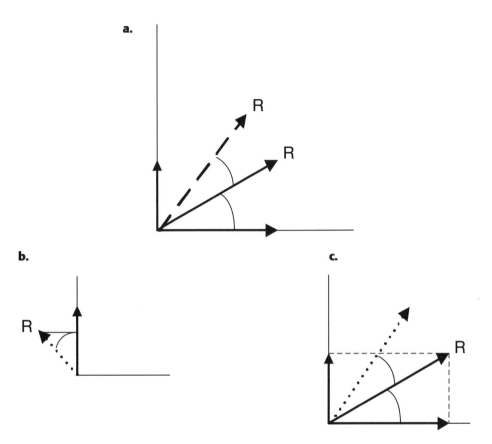

FIGURE 3.8 a) Rotation of the vector of magnitude R through an angle β. b) The rotation of X component through angle β. c) The rotation of Y component through angle β.

The rotated vector X and Y components are computed as

$$Y_{rot} = Y\cos(\beta) + X\sin(\beta)$$
$$X_{rot} = X\cos(\beta) - Y\sin(\beta)$$

FFT Algorithm

The best way to implement an FFT algorithm is to use the native code of the CPU instruction set, as there is an enormous amount of computation involved. But there is public domain software, such as FFTW (claimed to be the fastest Fourier Transform in the world), Scilab, and Grace have built-in FFT routines that you can freely download and use. You will see an implementation of FFT using the Grace algorithm in Appendix C, "Digial Filter Applications."

One of the main applications of FFT in digital signal processing is to compute the power spectrum of a random signal. The power spectrum describes the distribution of energy as a function of frequency, a measure of confidence in seeing a structure underneath a random signal.

RANDOM SIGNALS

Signals to some extent must be random if they are to convey any useful information. If a radio station were transmitting only one frequency, you would turn it off or switch to a new station. You learn something new only if a signal is unpredictable or has some randomness to it. Unwanted random interference, or noise, is present in all communication, but it also indicates the limit of useful information that can be carried over the signals. The statistical measure is the signal-to-noise ratio, and as an engineer, you always want to improve the signal-to-noise ratio.

For all practical purpose, signals or sequence may be considered as random bits of information, the average is the mean or dc value, the mean square is the average power, and the variance as a measure of how far away from the mean a given signal may be found. The probability associated with each allowed value may be defined as a probability distribution function, which is the running sum of all probabilities,

$$p_{x_n} = probablity\{x[n] = x_n\} = \sum_{-\infty}^{x_n} p_{x_n}$$

We would also like to find out if the successive sample values are truly independent or if there is a structure underneath the randomness. If signals are passed through a band-pass filter, we would expect the spectral energy to be closer to the central frequency. In any case, it is interesting to find the distribution of signal amplitude. This tells us the probability of finding a sample value at or within the range of certain values.

Amplitude Distribution

We start out with the basic experiment of a coin toss.

Having gone through n number of coin tosses and accumulating all the results up to now, you have a probability distribution function, which is the running integral of all the results. The sum must equal 1. It is a case of uniform distribution and there is an equal probability of occurrence for each event. But random signals and noise do not display such uniformity; the distribution is what is called Gaussian distribution.

Gauss observed the random errors in astronomical observations and concluded that the total errors in a given observation is part of a large number of individual errors, and he showed that the probability density function of the total errors is of the form

$$p(x) = \frac{1}{\sigma\sqrt{2\pi}} e^{\left(\frac{-(x-\bar{x})^2}{2\sigma^2}\right)}$$

The $\bar{x}$ is the mean value, σ^2 is the variance, σ is the standard deviation, and the constant $\sigma\sqrt{2\pi}$ is included to make the total area equal to 1.

Notice the Gaussian probability distribution—also called a normal distribution function—works only if you have repeated the experiment over and over again until infinity; otherwise, your result will only tend toward the underline probability distribution function.

The average power or mean square is

$$\overline{m^2} = E\{x[n]^2\} = \sum_{-\infty}^{\infty} x_n^2 P_{x_n}$$

The variance is similar to mean squared, but the mean is removed:

$$\sigma^2 = E\{(x[n]-m)^2\} = \sum_{-\infty}^{\infty} (x_n - m)^2 P_{x_n}$$

Autocorrelation

In digital signal processing we are more interested in finding out how successive sample values are related to each other, rather than the probability of finding a sample. We would like to see if the data is random or if there is a structure underneath, and we wish to identify its appropriate time series model. If we look at the coin toss experiment, the knowledge of previous results does not help us, as each outcome is independent, but if the samples go through a band-pass filter, we can predict that the sequence will bunch together around the center frequency of the filter, so a knowledge of recent history would help us predict the next value. The autocorrelation function characterizes such a time domain structure.

Mathematically, autocorrelation is the average product of the sequence with a time-shifted version of itself. If $x[1], x[2] \cdots x[K]$ is a sequence of events taken at equal space in time, the lag m autocorrelation function is defined as

$$r_m = \frac{\sum_{n=0}^{K-m-1} x[n] \times x[n+m]}{\sum_{n=1}^{K} x[n]^2}$$

The shift or the lag parameter m may have a value 1 through $K-1$. The method is simple and can be explained easily with a set of eight random data as tabulated in the top row of Table 3.5. Of course in real life, the experiment will have thousands of test results. We use a shift parameter $m = 3$. We lay out the numbers in columns shifted three places to the left and three places to the right, as shown in the rows of Table 3.5. We cross multiply each row with the original row, add together all products, and normalize by the sum of the squares of the original sequence.

TABLE 3.5 Autocorrelation of a Random Sequence

				19	−7	8	−1	3	−8	4	12				
m	Shifted sequence														r_k
−3	19	−7	8	−1	3	−8	4	12							−0.1017
−2		19	−7	8	−1	3	−8	4	12						0.1511
−1			19	−7	8	−1	3	−8	4	12					−0.2938
0				19	−7	8	−1	3	−8	4	12				1.000
1					19	−7	8	−1	3	−8	4	12			−0.2938
2						19	−7	8	−1	3	−8	4	12		0.1511
3							19	−7	8	−1	3	−8	4	12	−0.1017

Figure 3.9 is a stem plot of the computed values of r_k. To interpret these values, we perform a bigger experiment and perform autocorrelation on values obtained from a sinusoidal and a step function as shown in Figure 3.10. You can draw some important conclusions from autocorrelation as follows: the autocorrelation of a sinusoidal is another sinusoidal, the autocorrelation of a uniform sequence shows a tapered effect, but the random sequence can easily be identified with a peak value only at the center where the shift equals 0. Thus, the autocorrelation helps identify the structure underneath the time sequence.

Cross Correlation

Cross correlation describes the relationship between multiple processes; the method is similar to autocorrelation, but the product is performed between two different sequences: one is kept stationary while the other is time shifted. If the two processes are similar, there will be at least one maximum where the shift equals 0. Cross correlation is similar to convolution, but in convolution one sequence is time reversed.

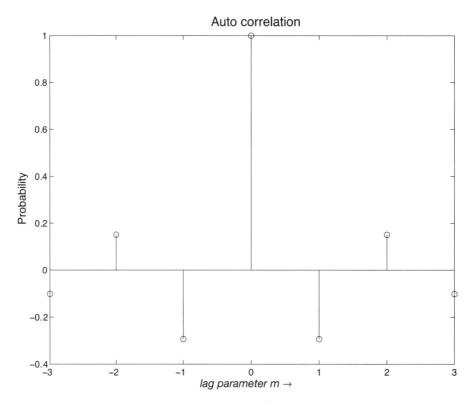

FIGURE 3.9 Autocorrelation of a random sequence of eight digits with a shift value of 3.

Power Spectrum

The power spectrum, or power spectral density, is the frequency domain counterpart of autocorrelation, in other words, power spectrum is the Fourier Transform of autocorrelation. It describes a signal's power (energy per unit time) falling within given frequency bins. It is not necessary to compute autocorrelation and then perform a Fourier Transform on it, the power spectrum may be computed directly by computing the square magnitude of the FFT. Figure 3.11 demonstrates the significance of the power spectrum as a mean to interpret the noisy data. Figure 3.11a is a time sequence of random data. Figure 3.11b is the power spectrum, indicating an even frequency spread between 0 to 500 Hz. Figure 3.11c is the same random data passed through an averaging filter. It is hard to see the effect of filtering, but Figure 3.11d shows the power spectrum of the filtered data, clearly demonstrating the filtering of high frequencies from the random data.

See Appendix A, "Matlab Tutorial," for the method of computing the power spectrum.

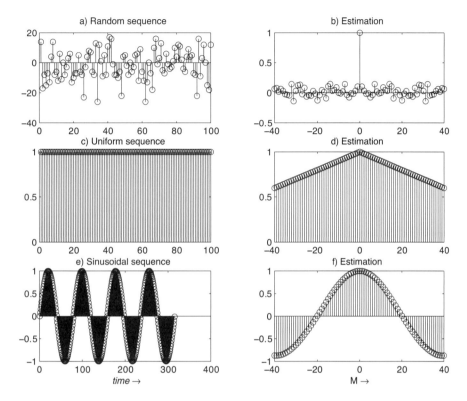

FIGURE 3.10 Autocorrelation. a) Random sequence. b) Autocorrelation of random sequence. c) Uniform sequence. d) Autocorrelation of uniform sequence. e) Sinusoidal sequence. f) Autocorrelation of sinusoidal sequence.

White Noise, Pink Noise, and 1/f Noise

A sequence of uncorrelated random numbers constitutes white noise, as it has the same distribution of power for all frequencies, meaning there is the same amount of power between 0 and 500 Hz, 500 and 1,000 Hz, or 20,000 and 20,500 Hz, etc. The power as a function of frequency is constant in white noise. To our ears, this seems very bright and harsh.

Pink noise on the other hand has an even distribution of power if the frequency is mapped in a logarithmic scale, meaning the same distribution of power for each octave, so the power between 0.5 Hz and 1 Hz is the same as between 5,000 Hz and 10,000 Hz. Since power is proportional to amplitude squared, the energy per Hz will decline at higher frequencies at the rate of about −3 dB per octave. The low-pass filtering of the white noise may make pink noise. To be more precise, the roll off should be −10 dB/decade, which is about 3.0102999 dB/octave.

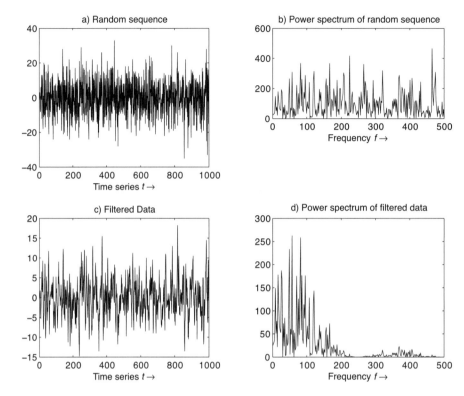

FIGURE 3.11 Power spectral density. a) The time sequencer of noisy data. b) Psd of the noisy data. c) Time sequence of filtered data. d) Psd of filtered data.

The use of the words pink and white may have come from the fact that if we mix light of different wavelengths evenly where the distribution is flat, the resulting light is white, so noise where $p(f)$ = constant, the power spectra is white noise, and if we mix visible light with different frequencies according to $1/f$ distribution, the resulting light may be pinkish (according to legend). Mixtures using other distributions should have different colors.

For a noise with $1/f$ power distribution, the volume decreases logarithmically with frequency. Consequently, it seems, our ears tell us that this is a "natural" even noise (it sounds like rushing water or ocean surf) and is quite relaxing. It's often used for ambience in electronic music (many equalizers and audio spectrum analyzers have built-in pink noise generators).

Thus, $1/f$ noise ("one-over-f noise", occasionally called "flicker noise") is a type of noise whose power spectra $P(f)$ as a function of the frequency f follows $P(f) = 1/f^a$, where the exponent a is very close to 1, as opposed to white noise where $P(f)$ = constant.

1/*f* noise appears in nature all over the place. The following is a description of a patent awarded for making an interesting use of the 1/*f* noise phenomenon, so, after all there is money to be made with DSP.

> An air conditioner for exchanging heat from a refrigerant to the outside and adjusting at least one of room temperature and humidity to desired temperature and humidity and having a 1/*f* fluctuation function for controlling an air supplying means for supplying conditioned air to the room so as to vary the air supplying amount in multiple levels corresponding to a designated reference air amount and irregularly. A reference air amount of the air supplying means can be designated and there is an air amount control for controlling the air supplying means corresponding to the designated reference air amount, and a fluctuation width designates a fluctuation width (volume of the air amount) corresponding to the air amount designated. When the 1/*f* fluctuation function is employed, the fundamental functions (coolness, warmness, and so forth) of the air conditioner are improved. In addition, the noise of the air conditioner is reduced and the comfort of the user is improved.
>
> Patent 5657640. Filed: Sep 14, 1995; Issued: Aug 19, 1997

SUMMARY

In this chapter, we have taken the long and cumbersome Fourier series and derived a short and concise form using Euler's identity. We removed the periodic dependency from the Fourier series and derived the Fourier Transform. Although the coefficients of the Fourier Transform no longer represent the amplitude of true sine and cosine functions, from practical point of view, it is sufficient to use them as relative magnitudes of frequencies. We also developed the method of Fast Fourier Transform that simplified the number of computations in Fourier Transform calculations.

We found that autocorrelation can help measure the randomness in a sequence of data, and the power spectrum describes a signal's power (energy per unit time) falling within a given frequency bins.

The discrete Fourier Transform:

$$X(n) = \sum_{k=0}^{K-1} x[k] \times e^{-j2\pi k \frac{n}{K}}$$

The inverse discrete Fourier Transform:

$$x[k] = \frac{1}{K}\sum_{k=0}^{K-1} X(n) \times e^{j2\pi k \frac{n}{K}}$$

The Fourier Transform:

$$X(f) = \int_{-\infty}^{\infty} x(t) \times e^{-j2\pi ft} \times dt$$

The inverse Fourier Transform:

$$x(t) = \int_{-\infty}^{\infty} X(f) \times (e^{j2\pi ft}) \times df$$

4 Solutions of Differential Equations

In This Chapter

- Linear Time Invariant Systems
- First Order Differential Equations
- The Convolution Process
- Second Order Differential Equations
- Natural Response
- Forced Excitations

So far we have studied Fourier analysis as a way to observe an event and analyze its constituent frequencies. Now we consider the processing of a signal: integrating, differentiating, smoothing, and filtering input signals to produce a desired outcome. Much of the processing in a digital signal processing refers to implementing solutions of differential equations, whether in a temperature control system or in speech analysis or vibration studies. The cause-and-effect relationship lends itself to the formulation of differential equations. We encounter them when we design analog electrical circuits and mechanical systems, and since digital signal processing has its roots in analog signal processing, it is imperative to understand the methodology of deriving the algorithms to solve differential equations.

In this chapter, we will devote our attention to the convolution process, as the method of solving a differential equation, while the other technique of Laplace Transforms will be deferred until the next chapter. If our goal is to design a simple integrator or differentiator, we need only derive a difference equation to be implemented as an iterative algorithm in a digital computer, but implementing a digital or analog filter requires deriving a closed form expression called a *transfer function*. A transfer function defines the output of a system as a function of frequency, indicating what frequencies will be suppressed while others are available without degradation. We begin our study with a refresher and establish the necessary mathematical foundation.

LINEAR TIME INVARIANT SYSTEMS

A system is considered linear if its output is directly proportional to its input, such as current and voltage relationships in an electrical system (for example, a nonlinear system is a relationship between current and power). The time invariance condition describes a system in which a delay in the input causes the same amount of delay in the output. The solutions that are presented in this chapter require the system to be linear and time invariant.

A linear system has an *additive* property and a *homogeneity* property. An additive system is one in which the response to a sum of inputs is the same as the sum of the individual responses, and a system is homogenous when the scaling of the input by some amount also results in the scaling of the output by the same amount (a sinusoidal input remains a sinusoidal output without affecting the frequency of the input signal; only the magnitude and phase may change). It should be noted that we will be dealing only with linear systems, and the differential equations will be linear differential equations.

A system will exhibit a certain response, depending upon the input energy applied to the system. If the response is due to the stored energy such as a charge on a capacitor or current in the inductor, it is the *natural response* of the system. But if the response is due to some external energy source, it is the *forced response* of the system.

The condition of a system being linear is not very strict. All we are asking is that, if the forced response of the system is being studied, there must not be any prior force present in the system, and if the natural response of the system is under study, the input force must have been removed after giving the initial push. The other additive and homogeneity property requires that the quantity under study must have a simple one-to-one relationship between the input and output of the system. A system exhibits an additive property if the output of several individual inputs is the same as the output of all the independent inputs summed together. For example, if the response of the input $2\sin(3t)$ is $\sin(3t + \phi_1)$ and $1.5\sin(4t)$ is $\sin(4t + \phi_2)$, then in a linear system the response of $2\sin(3t) + 1.5\sin(4t)$ would be $\sin(3t + \phi_1) + \sin(4t + \phi_2)$, as shown in Figure 4.1.

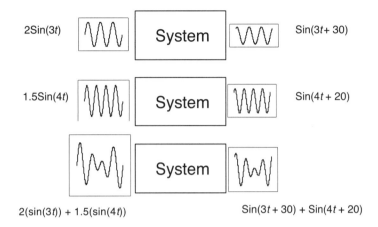

FIGURE 4.1 The additive property of a linear system; the combined output is the same as individual outputs summed together.

In other words, if the input is decomposed into several distinct excitations and the output is a superimposed result of each individual excitation, the system is additive in nature. The *time invariant* condition means if the input excitation is delayed by *t* amount of time, the output will be delayed by the same amount of time.

Differential Equations

A differential equation is formed when the function and the rate of change of the function (or the derivative) appear together in the same algebraic equation. The *order* of a differential equation is defined as the highest order of the derivative contained within it. A *first order differential equation* is formed in an electrical network if there is only one energy storage element in the circuit, such as a capacitor or an inductor. The current across the capacitor is proportional to the rate of change of the voltage applied, $i = C\frac{dv}{dt}$. Similarly, the voltage across an inductor is proportional to the rate of change of the current applied, $v = L\frac{di}{dt}$. (Resistors are not considered energy storage elements.) A *second order differential equation* is formed when there are two different energy storage elements in a circuit, such as a capacitor and inductor in series or parallel, resulting in a second derivative in the system. The current across a parallel combination of a capacitor and an inductor is $i = C\frac{d^2v}{dt^2} + \frac{1}{L}v$, while voltage across a series combination of a capacitor and an inductor is $v = L\frac{d^2i}{dt^2} + \frac{1}{C}i$.

Systems Response

A system that is capable of storing energy, such as an electrical circuit or mechanical system, responds to stimulus in two different ways, *natural response* and *forced response*. The behavior determined by the internal energy storage elements is the natural response, and the behavior determined by an external force is the forced response. Think about the energy stored when a spring is stretched. The spring may be forced to vibrate at any frequency by applying an external alternating force, but if the spring is stretched and released, it will vibrate at its characteristic frequency determined by the specific spring constant. The natural response is the response due to the stored energy being released at a natural pace. A system will always exhibit its natural response once the input excitation is removed from the system. Even if we don't remove it, as in the case of obtaining a forced response, we can think about the input as a series of impulses applied and derive a solution as if several natural responses occur sequentially. This is the basis for convolution and will be discussed later in the chapter. It is easier from an analysis point of view that we study the two responses independently and combine the result at our convenience at a later time. The system is guaranteed not to alter the behavior, since we are studying linear systems only.

Solutions of Differential Equations

Finding a response to an input excitation requires one to provide a homogenous solution as well as a particular solution. The *homogenous solution* is for the homogenous differential equation of the system, formed as a result of applying the basic laws of physics on the circuit components of the system, and it is devoid of an external force. It is also the natural response of the system, as it is the result of solving an equation when the input force is no longer in action. It is like seeing the residual effect due to the energy stored in the system. A *particular solution* of a differential equation is any function that satisfies the given differential equation. We will see that the particular solution is essentially the forced response of the system and the homogenous solution is the natural response of the system.

Linear Differential Equations

A linear differential equation is formed in a linear time invariant system as a result of meeting certain initial conditional criteria. For the forced response of the system, there must not be any force before the time $t < 0$, and for the natural response, the input force must be gone after time $t \geq 0$. These are the initial condition requirements for a differential equation to be linear. The initial condition for the natural response is also called *zero input condition*, and for the forced response, it is the *zero state condition*. The forced response of a system that already has an input force at

$t < 0$ creates a nonlinear differential equation, and does not apply to digital signal processing.

The homogenous solution and the particular solution can be derived separately for a linear system, and the final result may be obtained by a simple addition of the two solutions.

We can solve differential equations using the convolution method or the Laplace Transform method. Both are equally important in their own respect and are useful in different applications. Convolution is the primarily tool in image processing while Laplace Transform is used mainly in signal processing, such as speech and controls systems.

The Laplace Transform essentially converts a time domain signal into frequency domain and produces a response that is best suited for frequency analysis. Convolution on the other hand, is a time domain process and is mainly used in operations such as smoothing and filtering of input data.

The challenge in digital signal processing is to find a discrete time solution of a differential equation that has a counterpart continuous time solution. The theory of Laplace Transform is the basis for frequency domain analysis, and through that, the z Transform is derived, which forms the basis for a discrete sample solution. Similarly, convolution is the technique for solving time domain continuous time differential equations and there is a companion discrete time solution suitable for software algorithms of digital signal processing.

Filters, whether analog or digital, are implemented as close form expressions of differential equation solutions.

The General Form of a Differential Equation

An nth order differential equation is defined as

$$a_0 y + a_1 \frac{dy}{dt} + a_2 \frac{d^2 y}{dt^2} + \cdots + a_n \frac{d^n y}{dt^n} = x(t)$$

Where $x(t)$ and $y(t)$ represent the input and output respectively, and a_n are the constant coefficients describing the characteristics of the system elements. If the input excitation is removed (i.e., $x(t) = 0$), the resulting equation is a homogenous differential equation and requires only a general solution; otherwise, it is inhomogeneous and requires a particular solution. It should be noted that a complete solution is a general solution plus a particular solution.

In the scope of digital signal processing, only the solutions of the first and second order linear differential equation are considered. Other higher forms can be represented as cascaded or parallel combinations of simple first and second order equations. A solution of exponent form will always satisfy any order linear differential equation.

FIRST ORDER DIFFERENTIAL EQUATIONS

The general form of a differential equation with one energy storage element is given in Equation 4.1.

$$a_0 y + a_1 \frac{dy}{dt} = x \qquad (4.1)$$

Where y is the output and x is the input to the system.

Natural Response

Removing the input source at time $t = 0$, the input x becomes zero at $t \geq 0$, and the differential equation becomes

$$a_0 y + a_1 \frac{dy}{dt} = 0$$

We will assume a solution of exponential form

$$y = Ae^{kt}$$

$$\frac{dy}{dt} = kAe^{kt}$$

Selecting a value of $k = -\frac{a_0}{a_1}$ satisfies Equation 4.1.

The coefficient A can be solved using the initial condition of the system, and for that let's work through an example. The capacitor $C = 1uF$ in Figure 4.2 was initially charged with voltage V_0. At time $t = 0$, the switch was thrown in the position of the resistor $R = 1k$, essentially removing the power source from the system. For the circuit component as given, we will determine the voltage as a function of time after the capacitor is connected with the resistor and also find the discrete time solution of the network for the sampling rate of 10 samples per second.

Assume at $t = 0$ the switch was thrown towards the resistor. Applying the Kirchoff's voltage law on the RC network, we get a first order differential equation as a result of equating the voltages,

$$v_C(t) = v_R(t)$$

Using Kirchoff's current law of summing the current at a node, we get,

$$i_C(t) + i_R(t) = 0$$

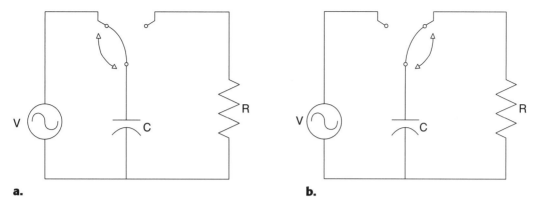

FIGURE 4.2 Circuit showing an RC network. a) The capacitor is connected to the power source. b) The power is switched off at $t \geq 0$.

The voltage and current across the resistor is

$$v_R = Ri_R \qquad i_R = \frac{v_R}{R}$$

The voltage and current across the capacitor is

$$v_c(t) = V_0 + \frac{1}{C}\int idt \qquad i_c = C\frac{dv_C}{dt}$$

We get a homogenous equation by summing the two currents

$$C\frac{dv_C}{dt} + \frac{v_C}{R} = 0$$

$$RC\frac{dv_C}{dt} + v_C = 0$$

Assuming v_C as an exponential function of time,

$$v_C = Ae^{kt} \qquad \frac{dv_C}{dt} = kAe^{kt}$$

Substituting the expressions into the homogenous solution leads us to Equation 4.2.

$$RCkAe^{kt} + Ae^{kt} = 0$$

$$(RCk+1)Ae^{kt} = 0$$

$$k = -\frac{1}{RC}$$

$$v_h = Ae^{-t/RC} \tag{4.2}$$

Equation 4.2 is the homogenous solution of a first order differential equation. To find the value of A, we apply the initial condition of the voltage $v = V_0$ at $t = 0$.

$$Ae^0 = V_0 \qquad A = V_0$$

Substituting the value of A and k in the expression, we obtained the relationship of the voltage as a function of time as the solution of homogenous differential equation, as shown in Equation 4.3.

$$v_H(t) = V_0 e^{-t/RC} \tag{4.3}$$

The particular solution is for the response due to the input force applied after the time $t = 0$, and for our example, we have 0 input force thus, Equation 4.4.

$$v_p = 0 \tag{4.4}$$

The complete solution is obtained by adding the homogenous solution (Equation 4.3) and the particular solution (Equation 4.4),

$$v_t = V_0 e^{-t/RC}$$

Figure 4.3 is the plot of the voltage as a function of time for an initial voltage of $1V$, $R = 1k$, and $C = 1uF$.

Forced Response

The natural response of a circuit will remain the same regardless of the input applied, but the response to an external source depends entirely on the input type. Although the study of a response to an arbitrary input is our goal, we derive the solution through the study of step response and impulse response. An impulse is considered a special case of the step (with a very short duration), and an arbitrary input could be considered as an input of a series of impulses very close to each other, albeit scaled by some magnitude and delayed by some time. If we know the

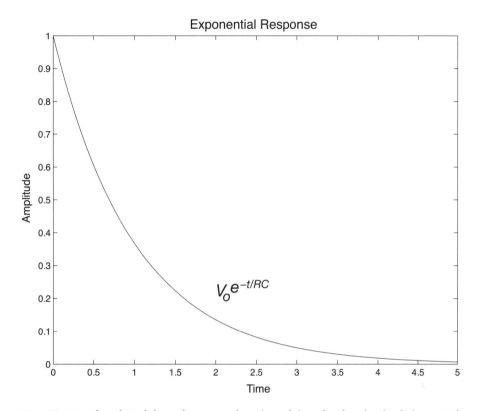

FIGURE 4.3 The plot of the voltage as a function of time for the circuit of Figure 4.2b.

response to a unit impulse, we can always find the response to an arbitrary input by considering the input as a series of impulses. Computing the cumulative affect of the previous responses and adding them to the output of the new response, we can find the complete response. The process is known as *convolution* and is simply a multiplication and addition operation; the multiplication scales the input to the unit impulse, and the addition takes into account the previous output. The frequency response is of special interest, as it helps us design circuits that act as frequency filters.

Step Response

Applying a constant current I at the instance $t = 0$ is synonymous to a step input. The switch in Figure 4.4, when thrown in the position of the capacitor, will let the current flow into the *RC* circuit. We can imagine the response conceptually that the charge on the capacitor will slowly build up until it reaches the level of the voltage that appears on the resistor R. Mathematically, we can calculate the time it takes for the capacitor to reach the final voltage ($v = RI$), and that would be the final response of the system for a step input.

Apply Kirchoff's Current Law on the circuit,

$$C\frac{dv}{dt} + \frac{1}{R}v = I$$

Where I indicates the constant current through the RC system.

The homogenous solution remains the same as described in Equation 4.2,

$$v_H = Ae^{-t/RC}$$

The particular solution is obtained by the new input condition of the constant current, and we get the following particular solution:

$$v_P = IR$$

You can verify the answer by placing the derivative $\frac{dv}{dt} = 0$ into the general differential equation. The complete solution, Equation 4.5, is obtained by adding the homogenous solution and the particular solution,

$$v = v_H + v_P$$
$$v = Ae^{-t/RC} + IR \tag{4.5}$$

The coefficient A is obtained from the initial condition

$$v(0) = 0$$
$$Ae^0 + IR = 0 \qquad A = -IR$$

Substituting the value of A into Equation 4.5, we obtain the voltage as a function of time, as in Equation 4.6.

$$v = IR(1 - e^{-t/RC}) \tag{4.6}$$

Figure 4.4b is the plot of Equation 4.6 showing the exponential rise of the voltage until it reaches the near full value of $v = RI$. The time constant T for an RC circuit is defined as the time it takes for the voltage to rise up to 0.63 of its full value and after the $4T$ the voltage reaches 99.98% of the full value.

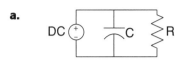

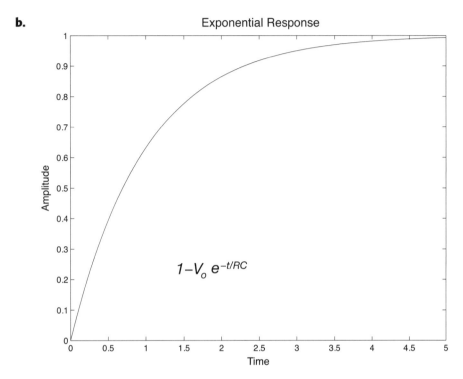

FIGURE 4.4 a) The RC network for Equation 4.6. b) The step response of the first order differential equation solution of Equation 4.6 for $I = 1$ and $R = 1$.

The Discrete Time Solution

Discrete time sampling takes inputs regularly at a predetermined interval. The term is implied in digital signal processing, as it takes a finite amount of time for a processor to acquire and process the data before the next sample is being obtained. If the sampling rate of the input signal is $1/\Delta t$ samples per second, the equivalent continuous time t for the kth sample is $k\Delta t$. To derive a discrete time solution of Equation 4.1, we use the method of backward differences,

$$RC \frac{y_k - y_{k-1}}{\Delta t} + y_k = x_k$$

112 Digital Signal Processing Fundamentals

Where x_k is the current input, y_k is the current output, and $y_{(k-1)}$ is the previous output, and the differential equation is simply a difference equation. Solving for y_k, we have

$$y_k = \frac{1}{1+\Delta t/RC} y_{k-1} + \frac{\Delta t/RC}{1+\Delta t/RC} x_k$$

If the Δt is sufficiently small, the term $(1 + \Delta t/RC)^{-1}$ may be approximated as

$$(1+\Delta t/RC)^{-1} \approx 1 - \Delta t/RC$$

Using a new term for $a_0 = \Delta t/RC$ and $b_1 = 1 - \Delta t/RC$, we get Equation 4.7.

$$y_k = a_0 x_k + b_1 y_{k-1} \qquad (4.7)$$

Using the input voltage $x = 1V$, $RC = 1$ and sampling period $\Delta t = 0.1$, we get

$$y_k = 0.1x + 0.9 y_{k-1}$$

The general solution described in Equation 4.7 is suitable for an iterative software algorithm for a digital computer implementation. Figure 4.5 shows the comparison between the discrete time solution and the continuous time solution of Equations 4.1 and 4.7.

Unit Impulse Response

By definition, a unit impulse is a short duration pulse with a total area equal to one. Figure 4.6 depicts such a function with a width of δ and a height of $\frac{1}{\delta}$ and is denoted by the symbol $\delta(t)$. It may not be possible to produce such a function in practice, but conceptually, it helps us formulate the response of a continuous function, for we can think of a continuous function of time as a series of impulses.

To find a solution of the homogenous equation for a unit impulse input, we need to find the amount of energy stored in the energy storage elements of the circuit (such as capacitors and inductors), due to the impulse applied. Consider the charge build up in a capacitor due to the current function whose area equals one,

$$\int i \, dt = CV = 1$$

$$V = \frac{1}{C}$$

Solutions of Differential Equations 113

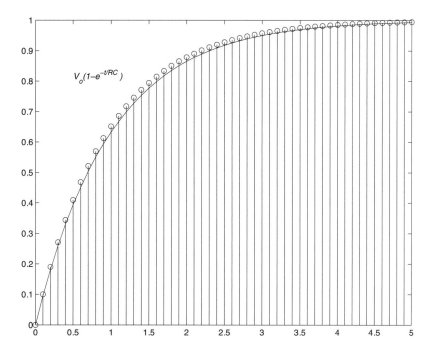

FIGURE 4.5 Comparison of the discrete time solution and the continuous time solution for the circuit of Figure 4.2.

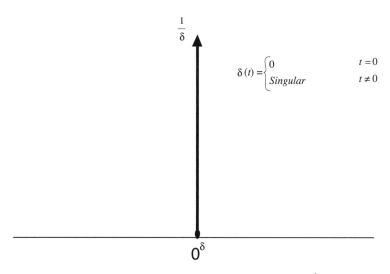

FIGURE 4.6 An impulse function of width δ and height $\frac{1}{\delta}$.

Similarly, the current stored in an inductor due to voltage spike function whose area equals one,

$$\int v\,dt = LV = 1$$

$$V = \frac{1}{L}$$

Suffice it to say that the initial condition produced by an impulse current on a capacitor is a voltage $V_0 = 1/C$, and the current stored in an inductor is $I_0 = 1/L$.

The requirement is to solve the solution of the following differential equation:

$$C\frac{dv}{dt} + \frac{1}{R}v = \delta$$

The particular solution is given as

$$v_p = \delta \quad \text{at } t = 0$$

For all practical purposes, the function $\delta = 0$ at time $t > 0$ make the particular solution

$$v_p = 0 \quad \text{at } t > 0$$

Using the previously mentioned initial conditions, we can obtain the complete response by adding the homogenous and particular solutions just like we did with the step input solution (Equation 4.8).

$$v = v_H + v_P$$
$$v = Ae^{-1t/RC} + \delta \quad \text{at } t = 0 \tag{4.8}$$

The coefficient A is obtained from the initial condition

$$v(0) = V_0 = 1/C$$
$$v = Ae^0 + \delta = 1/C \quad A = 1/C - \delta$$

Substituting the value of A into Equation 4.8, we obtain the voltage as a function of time,

$$v = (1/C - \delta)e^{-t/RC} + \delta \quad \text{at } t = 0$$

Substituting δ = 0 at time $t > 0$, we get the complete solution (or the response) to an impulse input in Equation 4.9.

$$v(t) = \frac{1}{C} e^{-t/RC} \tag{4.9}$$

Notice the similarities between the natural response (see Equation 4.2) and the impulse response of Equation 4.9. The comparison shows that the impulse response is essentially the same as the natural response of the system. The impulse function is being used only to visualize the amount of energy one can transfer in one go, without violating the initial condition of the zero input response. It may not be physically possible to provide such a force in reality, but it simplifies the mathematics. Once we determine the response of an impulse function, it is easier to derive the response of an arbitrary input function, and that will be explained in the convolution process section later in the chapter. Figure 4.7 is the plot of Equation 4.9 as $\Delta t \rightarrow 0$.

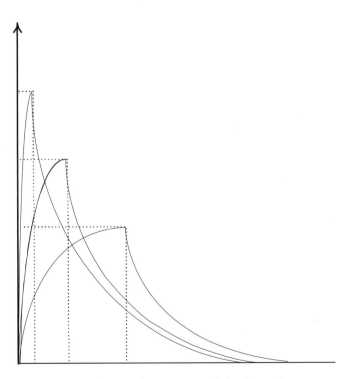

FIGURE 4.7 The impulse response of the first order differential equation as $\Delta t \rightarrow 0$.

Scaled Impulse Response

The impulse function doesn't have to be a unit area function. If the height is being halved, the area will be halved and the output would be simply half of what a unit impulse response is.

$$\frac{di_L}{dt}(0) = \frac{V_0}{V_L} = (A_1 S_1 + A_2 S_2)$$

$$V = \frac{0.5}{C}$$

$$v(t) = \frac{0.5}{C} e^{-t/RC}$$

In other words, if the input is scaled by a factor λ, the output would be

$$v(t) = \frac{\lambda}{C} e^{-t/RC}$$

Arbitrary Input and Convolution

It is easier to obtain a solution for differential equations if the input is a well-defined mathematical function such as a step, an impulse, or a sinusoidal input. Simply find the first and second derivative, plug in the values, and solve for the equations and get the output response. But an arbitrary input has no well-defined shape and form; a data acquisition system reading a channel value has no notion of the value being read. Events happen without set mathematical values. Take temperature and pressure for example. To design a control system for them, we must be able to predict the behavior of the system for an arbitrary input signal. In other words, find the response of the system to an arbitrary input.

One way to analyze such an input is to look through the window of an impulse. If we break down the input as if it is a series of scaled impulses, the job gets easier, as shown in Figure 4.8. We already know how to get the impulse response for an impulse function (see Equation 4.9 for an *RC* network). Now it is just a matter of finding the scaling factors and getting the scaled responses. Then simply add the individual responses (of course delayed by some time) and we have the desired outcome. This is convolution.

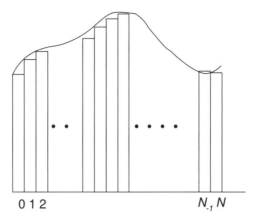

FIGURE 4.8 Approximating an input as a series of impulses.

THE CONVOLUTION PROCESS

The first step in convolution is to isolate the impulse from the rest of the input and then scale it. This creates a trail of unit impulse scaled by the input signals at specific instance of time, as shown in Figure 4.9. Without making it sound too complicated, if you think about it, the whole process is akin to simply taking the instantaneous values of the input signal at a specific interval of time. The value being acquired is the scaled unit impulse value, but of course delayed by the sampling interval. In this scheme, each new response will have a contribution from the previous response that we must take into consideration into the current output.

Discrete Time Convolution

Mathematically, we can express the operation of discrete time sampling as shown in Figure 4.9. A trail of areas delayed by the interval Δt. If t and $i_s(t)$ are the instantaneous sample time and sample value and t_n and $h(t_n)$ are the n_{th} sample time and unit impulse response, then the n_{th} delayed response is $i_s(t_n)h(t - t_n)\Delta t$. The one before that is $i_s(t_{n-0})h(t - t_{n-0})\Delta t$, all the way to the beginning $i_s(t_0)h(t - t_0)\Delta t$. We are going back in time and finding the response of the previous sample again but this time using the next part of the impulse response. Every time you multiply the current input with the current impulse response, you must add to it the previous

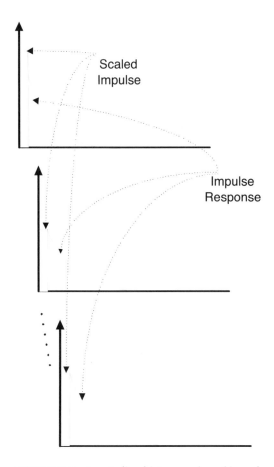

FIGURE 4.9 Input sliced into a series of impulse and response.

sample value multiplied by the delayed impulse response; see Figure 4.10 for a graphical description. Convolution processing is the mathematical operation of accumulating the current response plus all the previous responses; see Equation 4.10.

$$v(t_n) = i_s(t_0)h(t-t_0)\Delta t + i_s(t_1)h(t-t_1)\Delta t + \ldots + i_s(t_{n-1})h(t-t_{n-1})\Delta t$$

$$v(t_n) = \sum_{k=0}^{n-1} i_s(t_k)h(t-t_k)\Delta t \qquad (4.10)$$

Taking the limit as $\Delta t \to 0$, the summation in Equation 4.10 becomes an integral, as shown in Equation 4.11.

$$v(t) = \int_{t_0}^{t} i_s(t_k)h(t-t_k)dt \qquad t \geq 0 \qquad (4.11)$$

We can obtain the discrete time equivalent of Equation 4.10 by substituting the n_{th} impulse response and $t_k\Delta = k_{th}$ input sample, as shown in Equation 4.12. Notice the sign of the convolution operator $x(n)*h(n)$, a multiplication followed by the addition.

$$x(n)h(n) = \sum_{k=0}^{n} x(k)h(n-k) \qquad (4.12)$$

One disadvantage with Equation 4.12 is that the multiplication process is done over the entire array of input values, while in practice, the impulse response is usually short and most multiplications result in zeros. We can avoid this unnecessary multiplication by using the commutative property of the convolution as shown later in the section.

Equation 4.13 is the discrete time representation of a first order differential equation

$$y_k = a_0 x_k + b_1 y_{k-1} \qquad (4.13)$$

We can obtain a solution of this equation through the convolution process if the digitized version of the unit impulse response is provided. Assuming the discrete time impulse response is a series $h(0), h(1)...h(n-1), h(n)$ and the input samples are $x(0), x(1)...x(n-1), x(n)$, the convolution is simply a multiply and add operation, as shown in Equation 4.14.

$$y(k) = \sum_{i=0}^{k} h(i)x(k-i) \qquad (4.14)$$

NOTE

Although the impulse response may be infinite in length, after a while the response becomes negligible and for all practical purposes, values beyond the kth sample are treated as 0.

We can prove that Equation 4.14 is indeed a solution of Equation 4.13 (or Equation 4.1) with the following analogy.

Assuming the input is a unit impulse

$$x(k) = \delta(k) = \begin{cases} 1, k = 0 \\ 0, k \neq 0 \end{cases}$$

Substituting the unit impulse value $x(k)$ into Equation 4.13, we get the following series, shown in Equation 4.15.

$$y_0 = h_0 = a_0$$
$$y_1 = h_1 = a_0 b_1$$
$$y_2 = h_2 = a_0 b_1^2$$
$$\vdots$$
$$y_n = h_n = a_0 b_1^n \qquad (4.15)$$

But for an arbitrary input signal $x(k)$, we get

$$y_0 = a_0 x(0)$$
$$y_1 = a_0 x(1) + a_0 b_1 x(0)$$
$$y_2 = a_0 x(2) + a_0 b_1 x(1) + a_0 b_1^2 x(0)$$
$$\vdots$$
$$y(k) = a_0 x(k) + a_0 b_1 x(k-1) + \ldots a_0 b_1^k x(0)$$
$$y(k) = \sum_{i=0}^{k} h(i) x(k-i)$$

We can see that the coefficients of these equations match the impulse response of Equation 4.13, which is given by Equation 4.15.

Properties of Convolution

The convolution of two different sequences can be combined in different ways,

The Commutative Property

The order in which two sequences are convolved is not important. The following equality holds,

$$x(n) * h(n) = h(n) * x(n)$$

The Associative Property

If two systems with responses $h_1(n)$ and $h_2(n)$ are connected in series, an equivalent system is one that has a response equal to the convolution of $h_1(n)$ and $h_2(n)$.

$$x(n) * \{h_1(n) + h_2(n)\} = \{x(n) * h_1(n)\} + \{x(n) * h_2(n)\}$$

The Distributive Property

If two systems with responses $h_1(n)$ and $h_2(n)$ are connected in parallel, an equivalent system is one that has a response equal to the sum of $h_1(n)$ and $h_2(n)$.

$$x(n) * \{h_1(n) + h_2(n)\} = \{x(n) * h_1(n)\} + \{x(n) * h_2(n)\}$$

Graphical Representation of the Convolution Process

As described in Equation 4.14, the convolution is simply a multiply and add process, thus, for any discrete time input sequence $x(k)$ and the discrete impulse response sequence $h(k)$ of the system, the output sequence $y(k)$ may be computed using Equation 4.16,

$$y(k) = \sum_{i=0}^{k} h(i)x(k-i) \qquad (4.16)$$

You may recognize that we only need to perform the multiplication process for the range of numbers in which the impulse response $h(n)$ has nonzero values. Let's take an example of a system that has the following impulse response:

$$h(0) = 3$$
$$h(1) = 2$$
$$h(2) = 1$$
$$h(3) = 0$$
$$h(n) = 0$$

And the input sequence as shown in Figure 4.10.

Notice, the impulse response has zero values beyond the range $k > 2$. Thus, the convolution operation may be reduced as shown in Equation 4.17.

$$y(n) = \sum_{k=0}^{2} h(k) \times x(n-k) \qquad (4.17)$$

To illustrate the steps of discrete convolution, see Table 4.1. Write down the first sequence in the top row. Write down the inverse of the second sequence in subsequent rows shifted n times. Multiply and add all entries of the shifted sequence to the first sequence. Place the result in the last column. The entries in the last column are the result of the convolution.

The convolution operation forms the basis for the digital filtering technique that we will discuss in Chapters 6, "Filter Design" and 7, "Digital Filters."

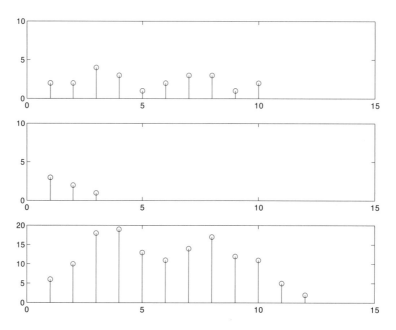

FIGURE 4.10 The plot showing the result of convolving sequence 2, 2, 4, 3, 1, 2, 3, 3, 1, 2 with 3, 2, 1.

TABLE 4.1 Convolution of Sequences, 2, 2, 4, 3, 1, 2, 3, 3, 1, 2 with 3, 2, 1

	–	–	2	2	4	3	1	2	3	3	1	2	–	–	
m	Shifted sequence														Summation
0	1	2	3												6
1		1	2	3											10
2			1	2	3										18
3				1	2	3									19
4					1	2	3								13
5						1	2	3							11
6							1	2	3						14
7								1	2	3					17
8									1	2	3				12
9										1	2	3			11
10											1	2	3		5
11												1	2	3	2

SECOND ORDER DIFFERENTIAL EQUATIONS

The simple exponential response of the first order differential equation was easy to visualize, but the second order differential equations are more complex in their responses, simply because there are two energy storage elements and their different possible combinations produce varying responses. Before we proceed with a full mathematical development, it would be helpful to create an intuitive feeling about the behavior of such systems in which two energy storage elements are in a loop, such as an inductor and a capacitor: one is capable of storing the current and the other is capable of storing the voltage.

A system as shown in the circuit of Figure 4.11 will serve the purpose for this example. We would like to see the inductive current as the response to the input voltage applied on the capacitor. Let's say the switch S1 on the capacitor C was originally connected to the voltage supply, letting the capacitor store a certain amount of charge. Once the capacitor is fully saturated, we throw the switch towards the inductor L, creating a loop between the inductor and the capacitor. The capacitor starts feeding the current to the inductor and the inductor starts building up the voltage across its terminals. The inductor gains the energy loss from the capacitor, but then the change of voltage across the inductor starts feeding the charge back into the capacitor. The charge gain by the capacitor back from the inductor is seen by the inductor as a current source and it starts building the voltage all over again. This back and forth yo-yo of energy loss and gain will last forever as long as we have ideal components.

The rise and fall of the charge on the capacitor and the current on the inductor is a sinusoidal function of time whose amplitude and wavelength depends upon the component values of the inductor and the capacitor of the circuit. To be more precise, the frequency of oscillation ω will be exactly equal to the value $\frac{1}{\sqrt{LC}}$ and is shown in Figure 4.11 for the circuit component of $L = 1$ and $C = 1$ with the initial voltage $V_0 = 1V$ and current $I_0 = 1$ amp.

Since there will be resistance to the current buildup in the inductor and the charge buildup in the capacitor, there will be a gradual loss of the energy and the sinusoid will die down eventually. We can expedite the loss by simply adding a resistive element to the circuit. This addition of a resistor will not only add to the exponential loss, but we will also see a decrement in the wavelength of the sinusoidal wave. The effect will be seen as if the sinusoid is being sandwiched between the two exponent curves, one rising from the negative and the other falling from the positive, both reaching the datum eventually, while squeezing the sinusoid along the way, as shown in Figure 4.12.

There is a chance that a fast exponent decay will not let the sinusoid ring at all and the whole thing will die down without showing any up and down motion.

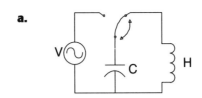

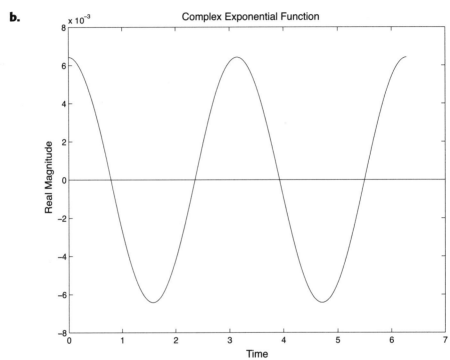

FIGURE 4.11 a) A circuit with two energy storage elements: an inductor and a capacitor. b) The sinusoid response of the two energy storage elements.

Otherwise, there will be a gradual decrement in the ringing and it will disappear altogether as time goes by. The exact phenomenon depends upon two factors, the quantities $\frac{R^2}{4L^2}$ and $\frac{1}{LC}$, if the resistor is placed in series, and $\frac{1}{4(RC)^2}$ and $\frac{1}{LC}$ if the resistor is placed in parallel (we will develop the mathematics later in the section, but for now, we will only use the terms). If $\frac{R^2}{4L^2}$ is greater than $\frac{1}{LC}$, we will not see any ringing at all; call it an over-damped condition. But if $\frac{R^2}{4L^2}$ is less than $\frac{1}{LC}$, there will be some ringing before reaching the finality; call it an under-damped condition. There is one critical value when $\frac{R^2}{4L^2}$ is just equal to $\frac{1}{LC}$, and this is the transition between being able to see a trough of the wave or not, a critically damped condition. The three responses are presented in Figure 4.12a, 4.12b, and 4.12c for a series *RLC* circuit and Figure 4.13a, 4.13b, and 4.13c for a parallel *RLC* circuit.

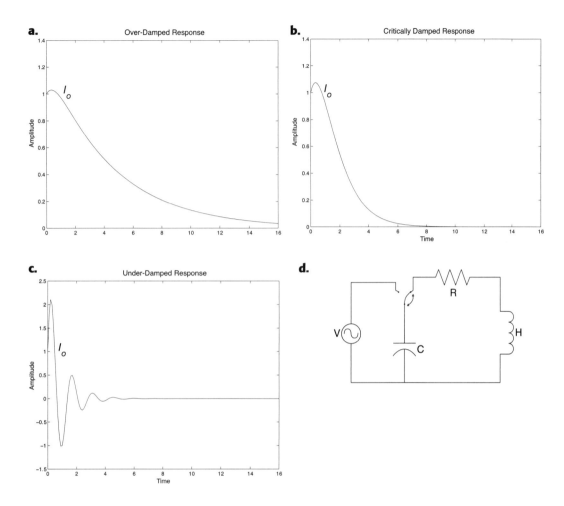

FIGURE 4.12 The series *RLC* circuit of second order differential equations. a) Over-damped response. b) Critically damped response. c) Under-damped response.

We have just discussed how inductors and capacitors form a resonant circuit and how adding a resistor puts a damper to the natural frequency. Our primary goal in this section is to study the output (i.e., the current on the inductor) in response to an input, the voltage on the capacitor. In this section, we will discuss the networks of electrical components and see how they form a system that effects an input excitation. We will analyze the system of resistors, inductors, and capacitors (*RLC*) in series as well as in parallel combination. Figure 4.12 is an example of a series circuit, and Figure 4.13 shows a parallel circuit. Our goal is to know how certain input frequencies are attenuated and others pass through without any change,

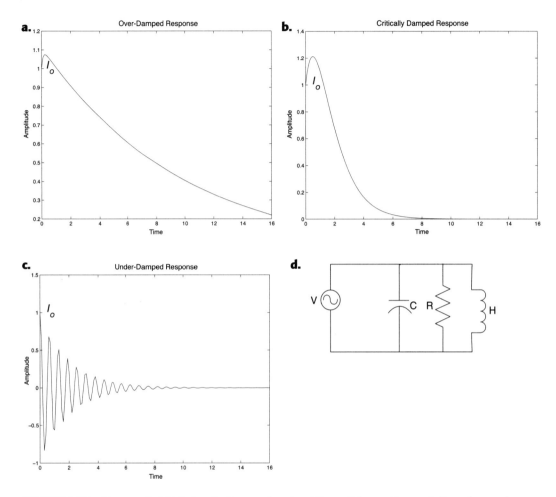

FIGURE 4.13 The parallel *RLC* circuit of second order differential equations. a) Over-damped response. b) Critically damped response. c) Under-damped response.

as we will be using them in our quest for designing filters. We begin with the mathematical formulation of the second order differential equation formed as a result of combining the three elements of an *RLC* circuit.

General Form of the Second Order Differential Equations

Equation 4.18 shows the general form of the second order differential equation with two energy storage elements,

$$a_0 y + a_1 \frac{dy}{dt} + a_2 \frac{d^2 y}{dt^2} = x(t) \tag{4.18}$$

Solutions of Differential Equations

The method of obtaining the solution is the same as that of the first order differential equation. We will derive the natural response of zero input condition giving us the homogenous solution and a forced response providing a particular solution (the forced response will be due to an external input excitation applied). The complete solution is obtained by adding the homogenous solution and the particular solution.

NATURAL RESPONSE

This is the response due to the internal stored energy only. There is no external input force, making $x(t)$ equal to zero. Thus, Equation 4.18 becomes,

$$a_0 y + a_1 \frac{dy}{dt} + a_2 \frac{d^2 y}{dt^2} = 0$$

Let's work through an example to find the solution, as we did with the first order differential equation. We will go through the series as well as parallel combination of *RLC* network simultaneously. The circuit of Figure 4.12 is a network of a resistor, capacitor, and an inductor (*RLC*) in series and Figure 4.13 is the network in parallel. In both cases, the capacitor was charged initially with the voltage V_0, before being switched to the network.

Following are the relationships between the current and voltages across different elements in the circuit.

$$i_C = C \frac{dv_C}{dt} \qquad i_R = \frac{v_R}{R} \qquad i_L = \frac{1}{L} \int v_L dt$$

$$v_C = \frac{1}{C} \int i \, dt \qquad v_R = R i_C \qquad v_L = L \frac{di_L}{dt}$$

At the time, the switch was thrown toward the capacitor. Kirchoff's current and voltage law describes the relationship for the parallel network as

$$i_C + i_R + i_L = 0$$

$$v_C = v_R = v_L$$

and for the series network as

$$v_C + v_R + v_L = 0$$

$$i_C = i_R = i_L$$

We get a second order differential equation as a result of summing the current in case of the parallel network, resulting in the following homogenous Equation 4.19.

$$C\frac{dv_C}{dt} + \frac{v_R}{R} + i_L = 0 \qquad (4.19)$$

Substituting the value of $v_R = L\frac{di_L}{dt}$ and $\frac{dv_C}{dt} = L\frac{d^2 i_L}{dt}$ in Equation 4.19, we get Equation 4.20.

$$\frac{d^2 i_L}{dt^2} + \frac{1}{RC}\frac{di_L}{dt} + \frac{1}{LC} i_L = 0 \qquad (4.20)$$

In case of the series network, summing the voltage provides Equation 4.21.

$$L\frac{di}{dt} + Ri + \frac{1}{C}\int i\, dt = 0 \qquad (4.21)$$

Taking the derivative of Equation 4.21, we get Equation 4.22.

$$\frac{d^2 i_L}{dt^2} + \frac{R}{L}\frac{di_L}{dt} + \frac{1}{LC} i_L = 0 \qquad (4.22)$$

Let's define the following terms.
   For series network,

$$\alpha = \frac{R}{2L}$$

And for parallel network,

$$\alpha = \frac{1}{2RC}$$

$$\omega_o = \frac{1}{\sqrt{LC}}$$

$$\alpha_d = \sqrt{\alpha^2 - \omega_o^2} \qquad \text{if } \alpha^2 < \omega_v^2$$

$$j\omega_d = j\sqrt{\omega_o^2 - \alpha^2} \qquad \text{if } \alpha^2 > \omega_v^2$$

Assuming the current i as an exponential function of time,

$$i_L = Ae^{st}$$

$$\frac{di}{dt} = sAe^{st}$$

$$\frac{d^2i}{dt^2} = s^2Ae^{st}$$

Substituting these expressions into the homogenous solution of Equations 4.20 and 4.22, we get Equation 4.23.

$$s^2Ae^{st} + s2\alpha Ae^{st} + \omega^2 Ae^{st} = 0 \qquad (4.23)$$

The two solutions for s are shown in Equation 4.24.

$$s_1 = -\alpha + \sqrt{\alpha^2 - \omega_o^2} \qquad s_2 = -\alpha - \sqrt{\alpha^2 - \omega_o^2} \qquad (4.24)$$

Both s_1 and s_2 satisfy Equation 4.23, and we get the solution describing the current on the inductor as a function of time in Equations 4.25 and 4.26.

$$i_L = A_1 e^{s_1 t} + A_2 e^{s_2 t} \qquad (4.25)$$

$$i_L = A_1(e^{-\alpha t} \times e^{(\sqrt{\alpha^2 - \omega_o^2})t}) + A_2(e^{-\alpha t} \times e^{-(\sqrt{\alpha^2 - \omega_o^2})t}) \qquad (4.26)$$

Where the value of A_1 and A_2 are to be obtained from the initial conditions.

The value $\alpha = \frac{1}{2RC}$ for a parallel network and $\alpha = \frac{R}{2L}$ for a series network determines the exponential decay, the value $\omega_v = \frac{1}{\sqrt{LC}}$ determines the original frequency of the sinusoid, and the value $\omega_d = \sqrt{\alpha^2 - \omega_0^2}$ determines the modified damped frequency of the sinusoid due to the presence of a resistive element in the circuit. The nature of the damped frequency ω_d suggests three possible answers:

- Roots real and distinct if $\alpha^2 > \omega_0^2$
- Roots real and equal if $\alpha^2 = \omega_0^2$
- Roots are complex if $\alpha^2 < \omega_0^2$

The two exponents of Equation 4.26 have an implied decay rate that primarily depends upon the value of α. The value $\alpha = \frac{R}{2L}$ is for the resistor in series, and for the resistor in parallel, $\alpha = \frac{1}{2RC}$. The quantity under square root ($\sqrt{\alpha^2 - \omega_0^2}$) needs further investigation. It should be noticed that the root is real for an over-damped circuit, since by definition it means $\frac{R^2}{4L^2}$ or ($\frac{1}{4RC^2}$) is greater then $\frac{1}{LC}$. On the other hand, if $\frac{R^2}{4L^2}$ is less then $\frac{1}{LC}$, the terms inside the square root $\sqrt{\frac{R^2}{4L^2}-\frac{1}{LC}}$ become a negative number, an under-damped condition whose value can only be evaluated by interchanging the terms and multiplying it by the imaginary operator $j = \sqrt{-1}$. The term $j\sqrt{\frac{1}{LC}-\frac{R^2}{4L^2}}$ or $(j\sqrt{\omega_0^2-\alpha^2})$ is our new damped frequency of oscillation. The newer frequency $\omega_d = \sqrt{\frac{1}{LC}-\frac{R^2}{4L^2}}$ is less then the original $\omega_0 = \sqrt{\frac{1}{LC}}$ by a factor of $\frac{R^2}{4L^2}$. The third option of critically damped condition of $\alpha^2 = \omega_0^2$ merely indicates a transition from over-damped to under-damp condition.

Evaluating the Coefficients A_1 and A_2

The solution presented in Equation 4.25 requires one to evaluate the coefficients A_1 and A_2 based on the initial conditions (the voltage V_0 present on the capacitor and the current I_0 on the inductor at time $t = 0$). It should be noted that for the series combination of the RLC circuit, the current I_0 gradually builds up in time, but at the instance $t = 0$ there is no current. That means I_0 is 0 while in the parallel combination of the RLC circuit the least resistive path to the current is through the inductor, resulting in an instantaneous current I_0 on the inductor originating from the capacitor.

Thus, for the series circuit, the initial conditions are shown in Equation 4.27.

$$i_L = 0 = A_1 + A_2 \qquad t = 0$$

$$\frac{di_L}{dt}(0) = \frac{V_0}{V_L} = (A_1 S_1 + A_2 S_2)$$

$$A_1 = (\frac{1}{S_1 - S_2})\frac{V_0}{L} \qquad A_2 = (\frac{1}{S_2 - S_1})\frac{V_0}{L} \qquad (4.27)$$

And for the parallel RLC, see Equation 4.28.

$$i_L = I_0 = A_1 + A_2 \qquad t = 0$$

$$\frac{di_L}{dt}(0) = \frac{V_0}{L} = (A_1 s_1 + A_2 s_2) \qquad t = 0$$

$$A_1 = (\frac{1}{s_1 - s_2})(\frac{V_0}{L} - s_2 I_0) \qquad A_2 = (\frac{1}{s_2 - s_1})(\frac{V_0}{L} - s_1 I_0) \qquad (4.28)$$

Roots Real and Distinct

If $\alpha^2 < \omega_0^2$, then the values of $s_1 = -\alpha + \sqrt{\alpha^2 - \omega_o^2}$ and $s_2 = -\alpha - \sqrt{\alpha^2 - \omega_o^2}$ become real, and the result is the sum of two exponent curves with no ringing of the sinusoid (an over-damped condition). Substituting the value $\omega_d = \sqrt{\alpha^2 - \omega_0^2}$ in s_1 and s_2, the two coefficients for the series RLC circuit are reduced to

$$A_1 = \frac{V_0}{2L\omega_d} \qquad A_2 = -\frac{V_0}{2L\omega_d}$$

And the current on the inductor is defined in Equation 4.29.

$$i_L = \frac{V_0}{2L\omega_d}(e^{-\alpha t} \times e^{(\sqrt{\alpha^2 - \omega_o^2})t}) - \frac{V_0}{2L\omega_d}(e^{-\alpha t} \times e^{-(\sqrt{\alpha^2 - \omega_o^2})t}) \qquad (4.29)$$

The output i_L of Equation 4.29 is plotted as a function of time in Figure 4.14 for the following component values in the series RLC circuit of Figure 4.12

$$R = 1\Omega \qquad C = 0.5F \qquad L = 0.1H \qquad V_0 = 1V$$

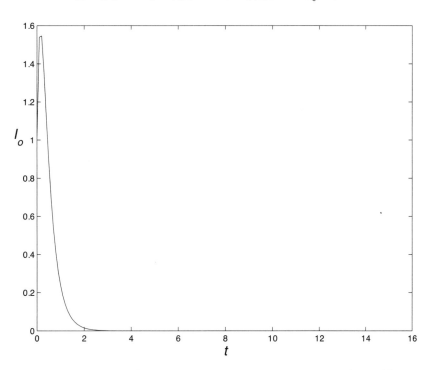

FIGURE 4.14 Output of $R = 1\Omega$, $C = 0.5F$, $L = 0.1H$, $V_0 = 1V$ produces this response.

Similarly for the parallel *RLC* network, the current on the inductor can be defined by substituting the constants of Equation 4.28 as shown in Equation 4.30.

$$i_L = (\frac{1}{s_1 - s_2})(\frac{V_0}{L} - s_2 I_0)(e^{-\sigma t} \times e^{(\sqrt{\alpha^2 - \omega_o^2})t}) + (\frac{1}{s_2 - s_1})(\frac{V_0}{L} - s_1 I_0)(e^{-\sigma t} \times e^{-(\sqrt{\alpha^2 - \omega_o^2})t}) \quad (4.30)$$

The output i_L of Equation 4.30 is shown in Figure 4.15 as a function of time for the following component values:

$$R = 1\Omega \quad C = 0.5F \quad L = 5H \quad V_0 = 1V \quad I_0 = 1A$$

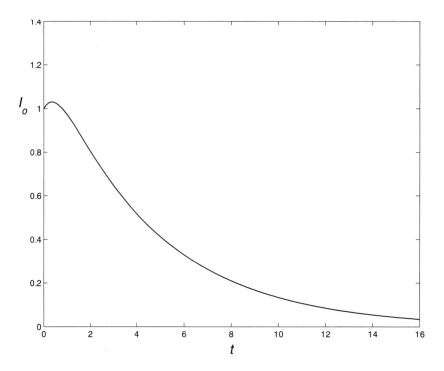

FIGURE 4.15 The over-damped exponent curve of the second order differential equation.

As you can see, the algebra gets very involved, but you can use Matlab to solve the equations. The figures were drawn using Matlab scripts, and the examples are presented in Appendix A, "Matlab Tutorial."

Roots Real and Equal

The value of $\alpha^2 < \omega_0^2$ is the critically damped condition, which is essentially a borderline condition where the system is just about to oscillate but not quite (a critically damped condition). By substitution, one finds that Ate^{kt} is also a solution, thus, the combined solution is shown in Equation 4.31.

$$i_L = (A_1 t + A_2) e^{-\sigma t} \tag{4.31}$$

Roots real and equal should not be a design consideration, as it creates a very unstable circuit. It is only a mathematical probability that the parameters are exactly equal, but it has no engineering significance. The constants A_1 and A_2 of Equation 4.31 are evaluated for the initial conditions of I_0 and V_0 as follows:

$$I_0 = A_2 \qquad t = 0$$

$$\frac{di_L}{dt} = \frac{V_0}{L} = A_1(e^{-\sigma t}) - \alpha(A_1 t + A_2) e^{-\alpha t}$$

$$\frac{V_0}{L} = A_1 - \alpha I_0 \qquad t = 0$$

$$A_1 = \frac{V_0}{L} + \alpha I_0$$

$$A_2 = I_0$$

For the parallel RLC circuit, the current on the inductor i_L is defined as in Equation 4.32.

$$i_L = (\frac{V_0}{L} + \alpha I_c) t e^{-\alpha t} + I_c e^{-\sigma t} \tag{4.32}$$

For the series RLC circuit $I_c = 0$, the current is defined as in Equation 4.33.

$$i_L = (\frac{V_0}{L}) t e^{-\alpha t} \tag{4.33}$$

The output i_L of Equation 4.32 for the parallel RLC circuit is shown in Figure 4.16 as a function of time for the following component values:

$$R = .5\Omega \qquad C = 1F \qquad L = 1H \qquad V_0 = 1V \qquad I = 1amp$$

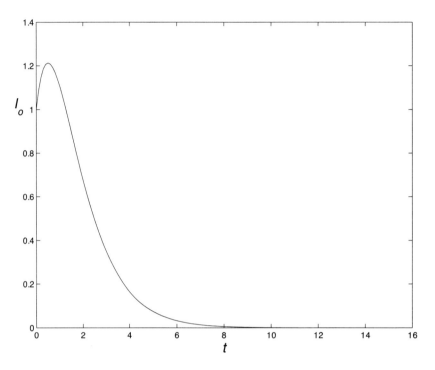

FIGURE 4.16 The output current as a function of time for a parallel circuit critically damped.

The output i_L for series RLC circuit of Equation 4.33 is shown in Figure 4.17 as a function of time for the following component values:

$$R = 2\Omega \quad C = 1F \quad L = 1H \quad V_0 = 1V \quad I = 1 amp$$

Roots Complex

If $\alpha^2 < \omega_0^2$, then the square root becomes a complex number, requiring you to interchange the terms ω_0^2 and α^2 by multiplying the root with the operator $j = \sqrt{-1}$. The two exponents give us the following solution:

$$i_L = A_1(e^{-\sigma t} \times e^{j(\sqrt{\omega_o^2 - \alpha^2})t}) - A_2(e^{-\sigma t} \times e^{-j(\sqrt{\omega_o^2 - \alpha^2})t})$$

This is the under-damped condition indicating two complex frequencies of oscillation $\pm j\sqrt{\omega_o^2 - \alpha^2}$, one rotating in a clockwise direction and the other counter-

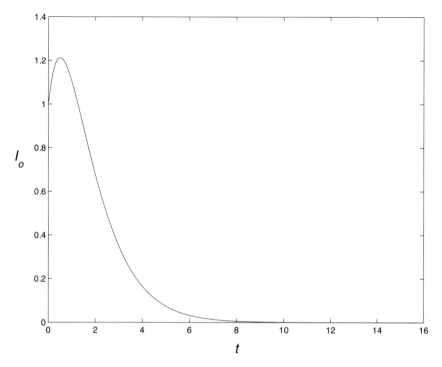

FIGURE 4.17 The output current as a function of time for a series critical damped circuit.

clockwise (see Figure 2.1, which describes two complex conjugate waves). The new wavelength is less than the original frequency ω_0 by a factor of α. With each frequency, there is an exponential decay multiplier $A_1(e^{-\sigma t})$ and $A_2(e^{-\sigma t})$. The system will respond with a sinusoid that will soon die down with an exponential decay rate of $e^{-\sigma t}$.

Using the identity $e^{j\omega t} = \cos(\omega t) + j\sin(\omega t)$ and substituting the value of A_1 and A_2 as described in Equation 4.28 for the parallel RLC circuit, we get the result shown in Equation 4.34 describing the current on the inductor as a function of time:

$$i_L = \frac{V_0}{L\omega_d} e^{-\alpha t} \sin\omega_d t + I_0 \frac{\omega_0}{\omega_d} e^{-\alpha t} \cos(\omega_d t - \phi)$$

$$\phi = \sin^{-1} \frac{\alpha}{\omega_0} \qquad (4.34)$$

Figure 4.18 is the plot of the current as a function of time for the following circuit component for the parallel RLC circuit of Figure 4.13.

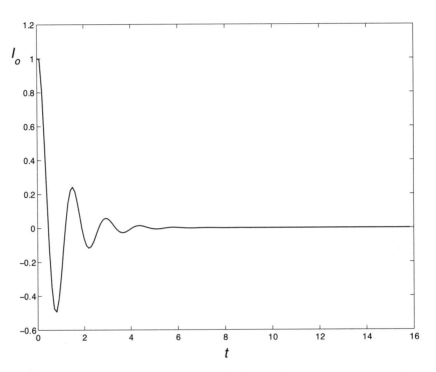

FIGURE 4.18 The plot of Equation 4.34, indicating the response of an underdamped parallel RLC circuit.

$$R = 2\Omega \quad C = 0.05F \quad L = 1H \quad V_0 = 1V$$

For the series RLC circuit, the result is shown in Equation 4.35.

$$i_L = \frac{I_0 \omega_0}{\omega_d} e^{-\alpha t} \cos(\omega_d t + \phi) - \frac{V_0 \omega_0^2 C}{\omega_d} e^{-\alpha t} \sin\omega_d t \qquad (4.35)$$

Figure 4.19 is the plot of the current as a function of time for the following circuit component for the series RLC circuit of Figure 4.12. The initial current I_0 in Equation 4.35 is considered 0, as the inductors resist change of current.

$$R = 2\Omega \quad C = 1F \quad L = 0.05H \quad V_0 = 1V$$

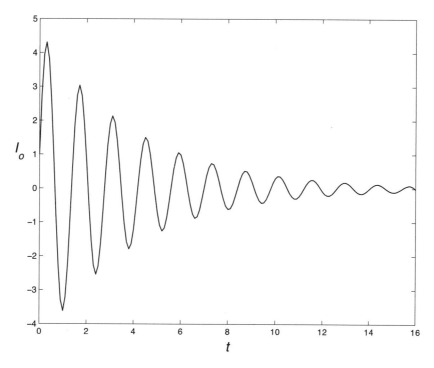

FIGURE 4.19 The plot of Equation 4.34, indicating the response of an under-damped series *RLC* circuit.

FORCED EXCITATIONS

The natural response discussed in the previous section was obtained for the zero input condition, since the input force was removed after the time $t >= 0$. Now we consider the zero state condition where an input force is applied after the time $t >= 0$. We begin with the discussion of the step and impulse response and then derive the response to an arbitrary input using the convolution process, as we did with the first order differential equations.

Step Response

A constant current *I* of a unit magnitude applied to an *RLC* network as shown in Figure 4.13 may be considered a step input $u(t)$. The response to such an excitation should be considered independently for the three conditions, namely roots complex ($\alpha^2 < \omega_0^2$) the under-damped condition, roots real ($\alpha^2 > \omega_0^2$) the over-damped, and roots equal ($\alpha^2 = \omega_0^2$) the critically damped condition. The following analysis uses the parallel *RLC* network as shown in Figure 4.13.

Applying Kirchoff's current law to the circuit of Figure 4.13 for the unit step function $u(t)$, we get Equation 4.36.

$$\frac{d^2 i_L}{dt^2} + 2\alpha \frac{di_L}{dt} + \omega_0 i_L = u(t) \tag{4.36}$$

The particular solution is obtained from the new input condition of the constant current, and we get the solution shown in Equation 4.37.

$$i_P(t) = 1 \quad t \geq 0 \tag{4.37}$$

The general solution is obtained for the three damped conditions of roots real, roots complex, and roots equal by adding the homogenous solution to the particular solution

$$i(t) = i_H(t) + i_P(t)$$

Roots Real

We get the following general solution by combining the homogenous solution of Equation 4.25 and the particular solution of Equation 4.37.

$$i(t) = A_1 e^{s_1 t} + A_2 e^{s_2 t} + 1$$

To resolve the coefficients A_1 and A_2, we need two independent equations, which we can obtain by using the following initial conditions:

$$i_L(0) = 0 = A_1 + A_2 + 1 \qquad \frac{di_L}{dt}(0) = 0 = A_1 s_1 + A_2 s_2$$

$$A_1 = \frac{s_2}{s_1 - s_2} \qquad A_2 = \frac{-s_1}{s_1 - s_2}$$

The step response is shown in Equation 4.38.

$$i(t) = \left[\frac{s_2}{s_1 - s_2} e^{s_1 t} + \frac{-s_1}{s_1 - s_2} e^{s_2 t} + 1 \right] u(t) \tag{4.38}$$

The coefficients s_1 and s_2 are real for the roots real condition,

$$s_1 = -\alpha + \sqrt{\alpha^2 - \omega_o^2}$$
$$s_2 = -\alpha - \sqrt{\alpha^2 - \omega_o^2}$$

For the parallel RLC network,

$$\alpha = \frac{1}{2RC} \qquad \omega_o = \frac{1}{\sqrt{LC}}$$

Figure 4.20 is a plot of Equation 4.38 for the following circuit components of Figure 4.13.

$$R = 1\Omega \qquad C = 1.0F \qquad L = 1H \qquad V_0 = 1V$$

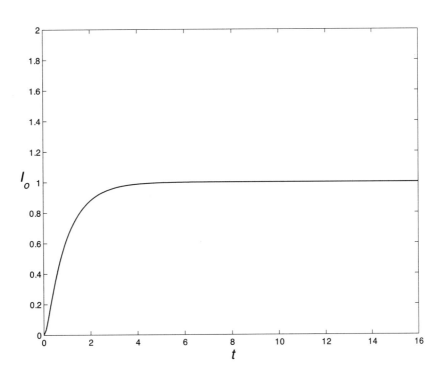

FIGURE 4.20 The inductive current of the parallel RLC circuit in response to a unit step input.

Roots Complex

The under-damped condition follows the same pattern as that of Equation 4.38, except now the constant s_1 and s_2 are complex conjugate numbers, since $\omega_v^2 > \alpha^2$. At this point, we will introduce the phasor equivalent of the complex numbers s_1 and s_2 to simplify the multiplication operation,

$$s_1 = -\alpha + j\sqrt{\omega_o^2 - \alpha^2} \qquad s_1 = -\alpha + j\omega_d \qquad |s_1| = \sqrt{\alpha^2 + \omega_d^2}$$
$$s_2 = -\alpha - j\sqrt{\omega_o^2 - \alpha^2} \qquad s_2 = -\alpha - j\omega_d \qquad |s_2| = \sqrt{\alpha^2 + \omega_d^2}$$

The polar representation of s_1 is defined as,

$$\frac{\pi}{2} + \phi = \tan^{-1}\frac{j\omega_d}{\alpha}$$

(The operator j adds $\frac{\pi}{2}$ to the phase angle.)

$$\omega_o = |s_1| = |s_2| = \sqrt{\alpha^2 + \omega_d^2}$$

$$s_1 = \omega_o e^{j(\frac{\pi}{2}+\phi)}$$

$$s_2 = \omega_o e^{-j(\frac{\pi}{2}+\phi)}$$

Substituting the complex values of s_1 and s_2 into Equation 4.38, we get the following simplified term resulting in Equation 4.39.

$$i(t) = \left[\frac{\omega_o}{s_1 - s_2}(e^{-j(\frac{\pi}{2}-\phi)}e^{(-\alpha+j\omega_d)t} - e^{j(\frac{\pi}{2}+\phi)}e^{-(-\alpha+j\omega_d)t}) + 1\right]u(t)$$

$$i(t) = \left[\frac{\omega_o e^{-\alpha t}}{s_1 - s_2}(e^{(j\omega_d - \frac{\pi}{2}-\phi)t} - e^{-(j\omega_d - \frac{\pi}{2}-\phi)t}) + 1\right]u(t)$$

$$i(t) = \left[\frac{\omega_o e^{-\alpha t}}{\omega_d}(\sin(\omega_d - \frac{\pi}{2} - \phi) + 1\right]u(t)$$

$$i_L(t) = \left[\frac{\omega_o e^{-\alpha t}}{\omega_d}(1 - \cos(\omega_d - \phi))\right]u(t) \qquad (4.39)$$

The graph in Figure 4.21 shows the under-damped frequency response to the unit step function of Equation 4.39 for the following circuit components:

$$R = 1\Omega \quad C = 4F \quad L = 1H \quad V_0 = 1V$$

It should be noticed that the exponentially decaying sinusoid reaches the value of 1 as time progresses.

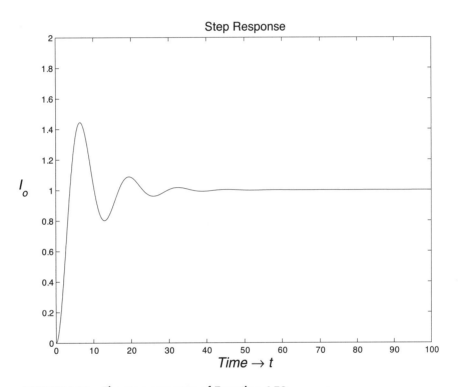

FIGURE 4.21 The step response of Equation 4.39.

The Complex Plane

The complex number representation of the coefficients s_1 and s_2 can best be described as a vector on a rectangular coordinate system. The x-axis is the damping factor α and the y-axis is the damped frequency ω_d, while ω_0 is the vector magnitude. Notice the damping factor α is always on the negative side of the quadrant for positive resistor values. This is the case with physical components in *RLC* circuits.

(The concept of negative resistance appears in some networks with feedback amplifiers that act like negative resistors, but this is beyond the scope of our analysis.) Figure 4.22 describes the rotation of such a vector on the complex plane with the magnitude ω_0 and the angle of rotation $\phi = \tan^{-1} \frac{\omega_d}{\alpha}$.

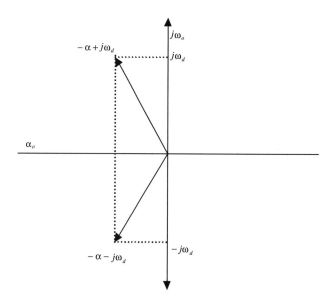

FIGURE 4.22 The complex conjugate vector with the damping coefficient.

The Quality Factor

The relationship between the damped frequency ω_d and the exponential damping factor α is obvious from the definition $\omega_d = \sqrt{\omega_o^2 - \alpha^2}$. The decrease in ω_d is in proportion to the increase in α. The ratio $\frac{\omega_0}{2\alpha}$ can be described as a quality factor in a system with a second order differential equation, such as the one being described in the series and parallel *RLC* circuit of Figures 4.12 and 4.13. To decrease damping, we must decrease α; a zero damping is an infinite Q, and that is the case for a true resonant circuit.

For a series *RLC* circuit, the Quality factor Q is defined as

$$Q = \frac{\omega_0}{2\alpha} = \frac{RC}{\sqrt{LC}} = R\sqrt{\frac{C}{L}}$$

And for a parallel circuit,

$$Q = \frac{\omega_0}{2\alpha} = \frac{R}{\sqrt{L/C}}$$

A damped resonant circuit can be described on the basis of the Quality factor Q as shown in Figure 4.23. A $Q < 1/2$ is over-damped, $Q = 1/2$ is critically damped, and $Q > 1/2$ is under-damped; $Q = \infty$ is a true resonant circuit with no loss of energy.

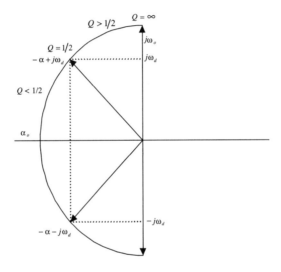

FIGURE 4.23 The Quality factor representation of the complex number vector $\omega_d = \sqrt{\omega_0^2 - \alpha^2}$ on a rectangular coordinate system.

Unit Impulse Response

Finding the response to an impulse function is simply a matter of determining the energy stored during a short impulse, as defined in the previous section of first order differential equations. To determine the initial conditions, first, we would analyze the effect of the impulse on the capacitor voltage. We know the area under the impulse; it is equal to 1 by definition. The charge injected due to an impulse will all be consumed by the capacitor to build up a voltage across the terminals.

$$\int \delta dt = CV = 1$$

$$V = \frac{1}{C}$$

The same voltage will appear across the inductor as the rate of change of the current i, as shown in Equation 4.40.

$$V = L\frac{di}{dt} \quad \text{or} \quad \frac{di_L}{dt}(0) = \frac{1}{LC} = \omega_0^2 \qquad (4.40)$$

The particular solution is given as:

$$i_P = \delta \text{ at } t = 0$$

The homogenous solution is given as:

$$i_H = A_1 e^{s_1 t} + A_2 e^{s_2 t} + \delta$$

The complete solution is the homogenous solution plus the particular solution:

$$i(t) = A_1 e^{s_1 t} + A_2 e^{s_2 t}$$

For all practical purposes, the function $\delta = 0$ at time $t > 0$ makes the particular solution

$$i_P = 0 \text{ at } t > 0$$

We solve simultaneous equations to obtain the coefficients A_1 and A_2,

$$i(0) = 0 = A_1 e^0 + A_2 e^0 \qquad A_1 = -A_2$$

$$\frac{di_L}{dt}(0) = \omega_0^2 = A_1 s_1 e^0 + A_2 s_2 e^0$$

$$A_1 = \frac{\omega_0^2}{(s_2 - s_1)} \qquad A_2 = \frac{-\omega_0^2}{(s_2 - s_1)}$$

The current across the inductor in the series RLC circuit for the under-damped case is therefore shown in Equation 4.41.

$$i_L(t) = \left[\frac{\omega_0^2}{s_1 - s_2}(e^{s_1 t} - e^{s_2 t})\right] u(t)$$

$$i_L(t) = \frac{\omega_0^2}{(2 j \omega_d)}(e^{-\alpha t + j \omega_d t} - e^{\alpha t - j \omega_d t})$$

$$i_L(t) = \frac{\omega_o^2}{\omega_d} e^{-\alpha t} \sin \omega_d t \tag{4.41}$$

Equation 4.41 must be multiplied with the unit step function $u(t)$ for it to be valid for all time t, which merely indicates that at time $t < 0$, the function is 0. See Equation 4.42.

$$i_L(t) = u(t) \frac{\omega_o^2}{\omega_d} e^{-\alpha t} \sin \omega_d t \tag{4.42}$$

Figure 4.24 is the graph of the impulse response. Notice the similarities between natural response as shown in Equation 4.35 and impulse response of Equation 4.42 when we consider the initial current $I_0 = 0$, which is the case for the series RLC circuit

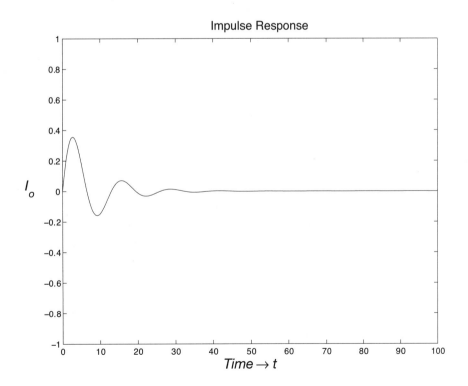

FIGURE 4.24 The impulse response of the second order differential equation for Equation 4.42.

Scaled Impulse Response

The response given in Equation 4.42 is for the unit impulse function whose area under the curve is equal to 1. Any other impulse that is a fraction λ of the unit impulse will produce the scaled response accordingly, as defined in the Equation 4.43:

$$i(t) = \lambda u(t) \frac{\omega_o^2}{\omega_d} e^{-\alpha t} \sin \omega_d t \qquad (4.43)$$

Response to an Arbitrary Input

The convolution process discussed in the previous section of the first order differential equation is also applicable to the second order system. The impulse response convolved with subsequent input to the system is the response to an arbitrary input. The input signal is treated as a series of delayed pulses, as shown in Figure 4.9, and the output is obtained by time-delayed addition of individual responses.

If $h(t - t_k)$ is the delayed impulse response and $i_s(t_k)$ is the current input, then the output is computed as the convolution sum shown in Equation 4.44.

$$v(t_n) = \sum_{k=0}^{n} i_s(t_k) h(t - t_k) \Delta \qquad (4.44)$$

We will be using the convolution method later on when we develop the algorithm to realize the digital filters.

SUMMARY

In this chapter, we discussed the input and output relationship of systems such as electrical circuits that are governed by differential equations. We established convolution as a method of solving the equations, and the discrete time convolution was discussed as a possible solution where a computer may be used for processing the input and output. We analyzed the series and parallel electrical networks for their varying responses by solving the corresponding differential equations. The goal in this chapter was to discuss the method of convolution as a solution of differential equations (it will form the basis of our filter design in the coming chapters) and to see the expected response of systems that are governed by differential equations.

5 Laplace Transforms and z-Transforms

In This Chapter

- The Laplace Transform
- Transfer Functions
- The z-Transform

We discussed in Chapter 4, "Solutions of Differential Equations," the method of convolution as a solution of differential equations. The problem with convolution is that it is a specific solution not a statement of system's behavior. It cannot be used as a model for the system. We need an expression that could describe the outcome of a system in terms of its dependent and independent variables, so design possibilities can be exploited. A plot of a frequency response is the most desirable picture in a digital or analog filter design, and for that, an algebraic form of a differential equation is required. We need to know what the output amplitude and phase change would be, given a set of input frequencies. The amplitude tells us how much gain or loss is impeded in the output and the phase tells us the fraction of a complete cycle elapsed from a specified reference point. In this chapter, we will solve these problems with the help of the Laplace Transform.

We discovered in the previous chapter that the exponentially decaying sinusoids, whether over-damped, under-damped, or critically damped are indeed solutions of linear differential equations, and the solution is only a matter of identifying the appropriate exponential coefficients. If the coefficients are real, the solution is simply an exponential decaying response, but if the coefficients are complex, we have a sinusoidal response with a damped frequency of oscillation. The damping factor determines the exponentially decaying amplitude. Similar to the Fourier Transform that identifies the component frequencies in a system, the Laplace Transform identifies the exponential decaying frequencies in a system. Just like a logarithm converts multiplication to addition, the Laplace Transform converts a differential into a multiplication and an integral into a division operation. Another consequence of the Laplace Transform is that the process eliminates the time dependency from the system, leaving behind only frequency as the independent variable.

THE LAPLACE TRANSFORM

If you recall the theory of Fourier Transforms discussed in Chapter 1, "Fourier Analysis," you would notice that we used the orthogonal property of sinusoidal function to identify the constituent frequencies of a given function. Broadening the concept and taking into consideration of the exponential damping factor, we introduce the integral in Equation 5.1 as the Laplace Transform of a function,

$$F(s) = \int_{0-}^{\infty} f(t) e^{-st} dt \qquad (5.1)$$

The quantity $s = \sigma + j\omega$ is essentially a combination of exponentially decaying factor and a sinusoidal function of frequency ω. If σ is considered 0 (meaning, insignificant damping effect), then the remaining exponent is the Fourier integral in the form of Euler's identity $e^{-j\omega t}$. In reality, as time goes by, all exponential decay settles down to insignificant values. In essence, the Laplace integral is only filling in for the missing transient response from the Fourier integral. Figure 5.1 is a map of the Laplace integral over the entire range of exponent decay σ and the frequency ω. The centerline is the Fourier integral, the left axis is the exponential decay, and the right axis is the exponential rise. The result of the Laplace integral is a function of frequency spectrum that identifies component frequencies in the system.

It is customary to indicate the Transform in capital letters. The integral is definite and the range is from –0 to infinity, thus eliminating the time factor, leaving behind a function that depends upon the complex variable s only. The range of integration is taken from –0 to incorporate the impulse function that starts at $t = 0$. Notice that if s is positive, the integral becomes infinite and that possibility needs to be avoided.

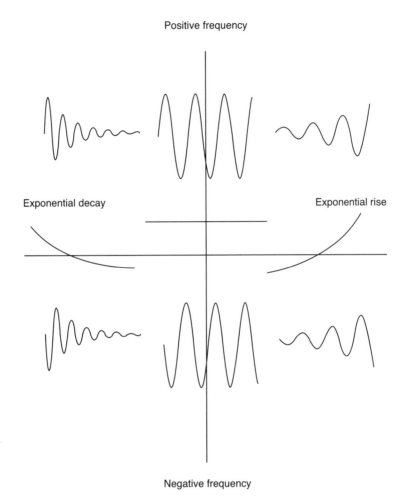

FIGURE 5.1 A map of Laplace integrals.

When we discuss the properties of the Laplace Transform, it becomes obvious that it provides linearity to a system. Think of the Transform as an operator that when applied, transforms differentials and integrals into simpler algebraic operations. Following are the properties of the Laplace Transform.

The Linearity Property of the Laplace Transform

The Transform of a sum of two functions is equal to the sum of individual Transforms; the linearity is evident from the following proof:

$$L[f_1(t)+f_2(t)] = \int_{0-}^{\infty} e^{-st}[f_1(t)+f_2(t)]dt$$

$$L[f_1(t)+f_2(t)] = \int_{0-}^{\infty} e^{-st}f_1(t)dt + \int_{0-}^{\infty} e^{-st}f_2(t)dt$$

$$L[f_1(t)+f_2(t)] = L[f_1(t)] + L[f_2(t)]$$

Some Useful Transformations

Following are some Transformations of some basic functions.

The *impulse function* has a nonzero value only at $t = 0$, and at $t = 0$, the exponent is evaluated as 1, and by definition the integral of the impulse function is a unit area, thus the Laplace Transform of an impulse function is 1, as shown in Equation 5.2.

$$L[\delta(t)] = \int_{0-}^{\infty} \delta(t)e^{-st}dt = \int_{0-}^{\infty} \delta(t)dt = 1 \qquad (5.2)$$

The *unit step function* is shown in Equation 5.3 (valid for positive s only).

$$L[u(t)] = \int_{0-}^{\infty} u(t)e^{-st}dt = \int_{0-}^{\infty} e^{-st}dt = \frac{e^{-st}}{s}\bigg|_{0-}^{\infty} = \frac{1}{s} \qquad (5.3)$$

The *unit step function with complex value s* is shown in Equation 5.4. If s is a complex value of the form $s = \alpha + j\omega$, the integral produces a complex quantity.

$$L[u(t)] = \int_{0-}^{\infty} u(t)e^{-(\alpha+j\omega)t}dt = \int_{0-}^{\infty} e^{-(\alpha+j\omega)t}dt = \frac{e^{-(\alpha+j\omega)t}}{-(\alpha+j\omega)}\bigg|_{0-}^{\infty} = \frac{1}{\alpha+j\omega} \qquad (5.4)$$

The *exponent function* is shown in Equation 5.5.

$$L[e^{at}] = \int_{0-}^{\infty} e^{at}e^{-st}dt = \int_{0-}^{\infty} e^{-(s-a)t}dt = \frac{e^{-(s-a)t}}{-(s-a)}\bigg|_{0-}^{\infty} = \frac{1}{s-a} \qquad (5.5)$$

The *cosine function* is shown in Equation 5.6.

$$L[\cos(\omega t)] = \int_{0-}^{\infty} \frac{1}{2}(e^{j\omega t} + e^{-j\omega t})e^{-st}dt = \frac{1}{2}\int_{0-}^{\infty} e^{-(s-j\omega)t}dt + \frac{1}{2}\int_{0-}^{\infty} e^{-(s-j\omega)t}dt$$

$$L[\cos(\omega t)] = \frac{1}{2}\frac{1}{s-j\omega} + \frac{1}{2}\frac{1}{s+j\omega} = \frac{s}{s^2+\omega^2} \quad (5.6)$$

The *sine function* is shown in Equation 5.7.

$$L[\sin(\omega t)] = \int_{0-}^{\infty} \frac{1}{2j}(e^{j\omega t} - e^{-j\omega t})e^{-st}dt = \frac{1}{2}\int_{0-}^{\infty} e^{-(s-j\omega)t}dt + \frac{1}{2}\int_{0-}^{\infty} e^{-(s-j\omega)t}dt$$

$$L[\sin(\omega t)] = \frac{1}{2j}\frac{1}{s-j\omega} + \frac{1}{2j}\frac{1}{s+j\omega} = \frac{\omega}{s^2+\omega^2} \quad (5.7)$$

The *exponentially decaying cosine function* is shown in Equation 5.8.

$$L[e^{\alpha t}\cos(\omega t)] = L[e^{\alpha t}\frac{1}{2}(e^{j\omega t} + e^{-j\omega t})e^{st}]$$

$$L[e^{\alpha t}\cos(\omega t)] = \frac{1}{2}\int_{0-}^{\infty} e^{(\alpha+j\omega)t}e^{-st} + \frac{1}{2}\int_{0-}^{\infty} e^{(\alpha-j\omega)t}e^{-st}$$

$$L[e^{\alpha t}\cos(\omega t)] = \frac{1}{2}\frac{1}{s-(\alpha+j\omega)} + \frac{1}{2}\frac{1}{s-(\alpha-j\omega)} = \frac{s-\alpha}{(s-\alpha)^2+\omega^2} \quad (5.8)$$

The *exponentially decaying sine function* is shown in Equation 5.9.

$$L[e^{\alpha t}\sin(\omega t)] = L[e^{\alpha t}\frac{1}{2j}(e^{j\omega t} - e^{-j\omega t})e^{st}]$$

$$L[e^{\alpha t}\sin(\omega t)] = \frac{1}{2j}\int_{0-}^{\infty} e^{(\alpha+j\omega)t}e^{-st} - \frac{1}{2j}\int_{0-}^{\infty} e^{(\alpha-j\omega)t}e^{-st}$$

$$L[e^{\alpha t}\sin(\omega t)] = \frac{1}{2j}\frac{1}{s-(\alpha+j\omega)} - \frac{1}{2j}\frac{1}{s-(\alpha-j\omega)} = \frac{\omega}{(s-\alpha)^2+\omega^2} \quad (5.9)$$

The Rules of Differentiation

The differential and integral rules are the two most important properties of the Laplace Transform that translate a differential equation into a simple algebraic linear equation. We can accomplish the Transform of a derivative by using the method of integration by parts, as shown in Equation 5.11.

$$L[\frac{df(t)}{dt}] = \int_{0-}^{\infty} \frac{df(t)}{dt} e^{-st} = e^{-st} f(t) \Big|_{0-}^{\infty} + s\int_{0-}^{\infty} f(t) e^{-st}$$

$$L[\frac{df(t)}{dt}] = sL[f(t)] \tag{5.10}$$

$$L[\frac{d^2 f(t)}{dt^2}] = s^2 L[f(t)] \tag{5.11}$$

Equations 5.10 and 5.11 simply state that the Transform of a derivative of a function is equal to the variable s times the Transform of the function itself. Notice, the Transform acts as an operator, which when applied, replaces the derivative with a linear expression.

The Rules of Integration

Similar to the derivation rule, the integration rule can be derived as shown in Equation 5.12.

$$L[\int_0^t f(t)] = \int_{0-}^{\infty} [\int_0^t f(t)] e^{-st} dt$$

$$L[\int_0^t f(t)] = [\int_0^t f(t) dt] \frac{e^{-st}}{-s} \Big|_{0-}^{\infty} - \frac{1}{-s} \int_{0-}^{\infty} f(t) e^{-st} dt$$

$$L[\int_0^t f(t)] = \frac{1}{s} L[f(t)] \tag{5.12}$$

The Laplace Transform of the integral of a function is $1/s$ times the Laplace Transform of the function itself, as shown in Equation 5.12.

Table 5.1 describes the Laplace Transform of some common functions. Notice some entries in the table have a multiplication factor of $u(t)$. This effectively provides a range of function values for $t < 0$, since the unit step function by definition has a value of 0 for $t < 0$ and a value of 1 for $t >= 0$.

We will use the differential and integral rules extensively in our analysis of systems that produce response to a stimulus in the form a differential equation.

The Inverse Laplace Transform

If you think of the Transform as creating a spectrum of the frequencies in a system, then think of the inverse as bringing all the frequencies back into a single function. It is like reverting to the time-based expression. The inverse operation is a complex integral in which all the frequencies in a system are summed together to recover the original time-based function; the integral in Equation 5.13 describes the inverse operation.

$$x(t) = \frac{1}{2\pi j} \int_{\alpha-j\infty}^{\alpha+j\infty} X(s)e^{st} ds \qquad (5.13)$$

In practice, however, a table of Transform pairs (such as Table 5.1) derived in Equations 5.2 through 5.9 are being used to identify the corresponding time-based function for a given Transformed expression.

Solving Differential Equations with the Laplace Transform

The differentiation and integration rules of the Laplace Transforms, derived in Equations 5.10, 5.11, and 5.12, transform an ordinary differential equation into an algebraic expression. The expression can be simplified and solved for the dependent and independent variables, but the solution is still an expression in the form of a Laplace Transform. If a true solution is desired, the Transformed output must be converted to the time-based function, either by using an inverse Laplace integral or using a table lookup such as Table 5.1 to identify the appropriate Transform pair.

You will soon realize that all the necessary design information is present in the Transformed output, such as the system's Gain and Phase change at different frequency input, etc. But for now, we would like to show that the Transform indeed provides a solution of differential equations. One advantage of the Laplace Transform method is that the Transform is simply a frequency dependent linear function, ideally suited for analyzing the system's response at different frequencies; in fact, the original answer is not as important as the Transform plot of the magnitude versus the frequency that can describe the system's behavior over the entire range of frequencies of interest. You will see the use of frequency response later.

Let's take the example of the impulse response of a series *RLC* circuit as described in Figure 5.2. We have already discussed this example in the previous chapter and solved the problem using the conventional method, and now we would like to use the method of substitution by the Laplace Transform.

TABLE 5.1 Laplace Transform of Some Common Functions

	$f(t)$	$F(s) = \int_{0^-}^{\infty} f(t) e^{-st} dt$
1	$\delta(t)$	1
2	$u(t)$	$\dfrac{1}{s}$
3	$e^{-\alpha t} u(t)$	$\dfrac{1}{s+\alpha}$
4	$t^n u(t)$	$\dfrac{n!}{s^{n+1}}$
5	$\sin(t) u(t)$	$\dfrac{\omega_0}{s^2 + \omega_0^2}$
6	$\cos(t) u(t)$	$\dfrac{s}{s^2 + \omega_0^2}$
7	$e^{\alpha t} \sin(\omega t)$	$\dfrac{\omega}{(s-\alpha)^2 + \omega^2}$
8	$e^{\alpha t} \cos(\omega t)$	$\dfrac{s-\alpha}{(s-\alpha)^2 + \omega^2}$
9	$a e^{-\alpha t} \cos \beta t + \dfrac{(b - a\alpha)}{\beta} e^{-\alpha t} \sin \beta t$	$\dfrac{as+b}{(s+\alpha)^2 + \beta^2}$

The relationships between the current and voltages across different elements in the circuits are

$$i_C = C \frac{dv_C}{dt} \qquad i_R = \frac{v_R}{R} \qquad i_L = \frac{1}{L} \int v_L dt$$

$$v_C = \frac{1}{C} \int i \, dt \qquad v_R = R i_C \qquad v_L = L \frac{di_L}{dt} \qquad \frac{dv_L}{dt} = L \frac{d^2 i_L}{dt^2}$$

The Kirchoff's current and voltage law describes the relationship for the series network as

$$i_C = i_R = i_L$$
$$v_C + v_R + v_L = 0$$

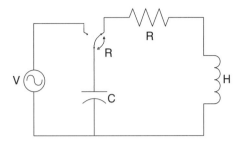

FIGURE 5.2 Series *RLC* circuit showing capacitor switched to the resistor.

The second order differential equation is formed because of summing the voltages. Let *h* be the impulse response, resulting in Equation 5.14.

$$L\frac{dh}{dt} + Rh + \frac{1}{C}\int hdt = \delta(t) \tag{5.14}$$

Conventionally, the Transform is expressed in bold letters; so let *H(s)* be the Transform of the impulse response. The integrals and differentials of Equation 5.14 are then reduced to the following algebraic form:

$$L[L\frac{dh}{dt}] = LsH(s)$$

$$L[Rh] = RH(s)$$

$$L[\frac{1}{C}\int hdt] = \frac{1}{Cs}H(s)$$

The Laplace Transform of the impulse input from Equation 5.2 gives us

$$L[\delta(t)] = 1$$

Substituting the Transforms into the terms of Equation 5.14, we get the following algebraic equations:

$$(Ls + R + \frac{1}{Cs})H(s) = 1$$

$$H(s) = \frac{R}{L}\frac{s}{s^2 + (R/L)s + 1/LC}$$

Substituting the coefficients with the standard notations,

$$\omega_0^2 = \frac{1}{LC} \qquad \alpha = \frac{R}{2L} \qquad \omega_d = \sqrt{\omega_0^2 - \alpha^2} \qquad \phi = \sin^{-1}\frac{\alpha}{\omega_0}$$

Solving for $H(s)$, we get Equation 5.15.

$$H(s) = \frac{R}{L}\frac{s}{(s^2 + 2\alpha s + \alpha^2) + (\omega_0^2 - \alpha^2)}$$

$$H(s) = \frac{R}{L}\frac{s}{(s+\alpha)^2 + (\omega_d^2)} \qquad (5.15)$$

Although we could perform the inverse Laplace Transform of Equation 5.15 to obtain the time domain function, with some manipulation, the result can be made to fit the pattern being described in entry 9 of Table 5.1 (Equation 5.16).

$$h(t) = \frac{R}{L}e^{-\alpha t}(\cos\omega_d t - \frac{\alpha}{\omega_d}\sin\omega_d t)$$

Using the trigonometric identity $\cos(x+y) = \cos x \cos y - \sin x \sin y$, we can rewrite the impulse response as shown in Equation 5.16.

$$h(t) = \frac{\omega_0 R}{\omega_d L}e^{-\alpha t}(\cos\omega_d t + \phi) \qquad (5.16)$$

The solution of Equation 5.14 obtained through the Laplace Transform matches the result of Equation 4.35 that was obtained through the conventional method.

It is obvious that the Laplace Transform could be used as an effective method of solving differential equations, but engineering design is not only about solving specific problems but modeling a system and analyzing the system's response to various input conditions. A Transfer function of a system is a mathematical model that describes the system's response to input frequencies and establishes a relationship between the input and output (also known as the *gain*) of the system. Even if a system is complex (having multiple stages of input and output), still we can perform the Transform on its subsystems and linearly add or multiply each subsystem at later stages to obtain the overall system response. That leads to the concepts of impedance and admittance as the Transform of the lowest level of subsystems, meaning the component level of inductors, capacitors, and resistors.

Impedance and Admittance

The lowest levels of subsystems in electrical networks are the individual resistors, capacitors, and inductors, and for a linear system, the overall system response is the same as the sum of the individual responses. Impedance and admittance are the transforms of basic elements of such subsystems. In electrical networks, the input and output are usually voltages or current. Impedance is defined as the ratio of the Laplace Transform of the output voltage and the input current, while admittance is defined as the ratio of the output current and input voltages.

When components are placed in series, their impedances are added, and for elements connected in parallel, the admittance is added. We have already seen that for resistance in series, the individual resistances are added, and for resistance in parallel, the admittance is added. The impedance and admittance are simply the input and output relationships of voltage and current, which is equally applicable to all elements including the resistors, capacitors, and inductors.

Figure 5.3a, b, and c describes the impedance and admittance of the three basic elements of an electrical network, namely, the resistors, capacitors, and inductors.

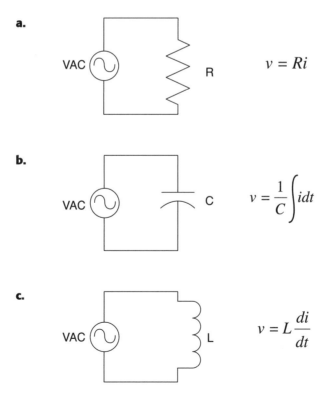

FIGURE 5.3 Impedance and admittance of a) resistors, b) capacitors, c) inductors.

The impedance Z of a resistor is the Laplace Transform of the output voltage and the input current, while the admittance G is the Laplace Transform of the output current and the input voltage.

$$Z_R = \frac{L[v]}{L[i]} = \frac{V}{I} = R \qquad G_R = \frac{L[i]}{L[v]} = \frac{1}{R}$$

Similarly, the input current and output voltage and output current and input voltage across capacitors and inductors are defined in terms of the Laplace Transform as

$$Z_L = \frac{L[v]}{L[i]} = \frac{V}{I} = sL \qquad G_L = \frac{L[i]}{L[v]} = \frac{1}{sL}$$

$$Z_C = \frac{L[v]}{L[i]} = \frac{V}{I} = \frac{1}{sC} \qquad G_C = \frac{L[i]}{L[v]} = sC$$

Applying the concept of impedance to the RC circuit of Figure 5.4a, we could obtain the response with the following derivative,

The current through the loop is

$$I = V_{in}/(Z_R + Z_C)$$

The voltage across the resistor R is

$$V_{out} = \frac{V_{in}}{Z_R + Z_C} Z_R = \frac{V_{in}}{1 + \frac{1}{sRC}}$$

The overall system response is therefore,

$$H(s) = \frac{V_{out}}{V_{in}} = \frac{1}{1 + 1/sRC}$$

$$H(s) = \frac{V_{out}}{V_{in}} = \frac{sRC}{sRC + 1}$$

As mentioned earlier, the quantity $s = \sigma + j\omega$ is a combination of exponentially decaying factor and a sinusoidal function of frequency ω. Ignoring the exponent decay factor as merely initial transient response contributor, we can formulate the steady state response as a function of frequency ω as

$$H(s) = \frac{V_{out}}{V_{in}} = \frac{j\omega RC}{j\omega RC + 1}$$

or

$$H(s) = \frac{V_{out}}{V_{in}} = \frac{j\omega RC}{1 + j\omega RC} \times \frac{1 - j\omega RC}{1 - j\omega RC} = \frac{j\omega RC + (\omega RC)^2}{1 + (\omega RC)^2}$$

Multiplying it with its complex conjugate and taking the square root gives the magnitude $|H(s)|$ (Equation 5.17).

$$|H(s)| = \sqrt{H(s) \times {}^*H(s)} = \sqrt{\frac{(\omega RC)^2 + j\omega RC}{1 + (\omega RC)^2} \times \frac{(\omega RC)^2 - j\omega RC}{1 + (\omega RC)^2}}$$

$$|H(s)| = \sqrt{H(s) \times {}^*H(s)} = \frac{\omega RC}{\sqrt{1 + (\omega RC)^2}} \tag{5.17}$$

A plot of $H(s)$ as the gain versus s (frequency variable) is presented in Figure 5.4b for the values of $C = 1$ and $R = 2$, and a glance at the plot shows that at high frequencies ($s \gg RC$) the gain is much higher compared to low frequencies ($s \ll RC$), essentially suppressing the low frequencies but letting the high frequencies pass through. In fact, at 0 frequency (meaning DC level) the output is essentially 0. Clearly, the circuit of Figure 5.4 is a high-pass filter. Such analysis of system's response is possible only through the Transfer function of a system.

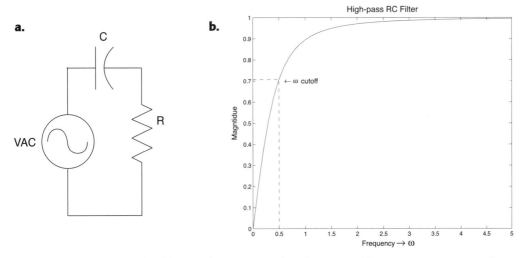

FIGURE 5.4 a) *RC* circuit with capacitor connected to the power. b) Step response across the resistor of the *RC* circuit using linear scale.

A better way of presenting the frequency plot is through a logarithmic scale, to highlight a large range of frequency values. Figure 5.5 is the plot of Equation 5.17, using the logarithmic scale for the *x*-axis.

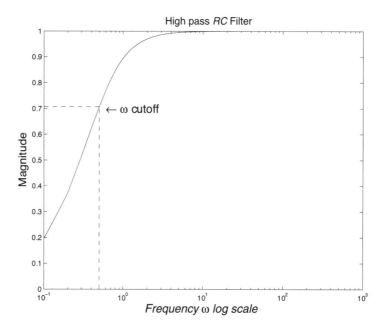

FIGURE 5.5 Step response of the circuit in Figure 5.4 using the logarithmic scale.

TRANSFER FUNCTIONS

The example of the high-pass filter presented in the previous section shows that the simplification of a differential equation using the Laplace Transform offers a great tool for analyzing a system's behavior. The algebraic relationship of input and output in the form of a system's response is ideal for describing a model for the system—especially, the frequency response of the system. We have already seen in the Fourier Transform that any input to a system can be described in terms of its constituent frequencies, and having a model of frequency response in the form of the Laplace Transform just about sums up everything in the nature that we need to analyze.

Although the impulse response developed in the previous section can be used for the frequency analysis, so that we can simply plot the magnitude as a function of frequency, we seek a uniform representation that is applicable to all systems. For now, we focus our attention on how a Transfer Function helps us in analyzing the frequency response of a system, and later in the section of poles and zeros and Bode plot, we will develop a uniform approach that is applicable to all types.

A Transfer Function of a system is the input and output relationship in the form of frequency response that is suitable for frequency analysis. In the context of the Laplace Transform, a system with input voltage $V_{in}(s)$ and output voltage $V_{out}(s)$ having an impulse response $H(s)$, has the following Transfer Function:

$$\frac{V_{out}(s)}{V_{in}(s)} = H(s)$$

The significance of the impulse response is that in the form of Laplace Transform, it provides a frequency dependent algebraic relationship, just what we needed to analyze the system's behavior at various frequencies. Notice the difference between the Fourier Transform and Laplace Transform: the complex variable $s = \sigma + j\omega$ in the Laplace Transform includes the exponential decaying factor as well as the natural frequency of the system, whereas the Fourier Transform has only the natural frequencies of the system.

It is true that after some time, all systems follow the frequency of the input signals and the original transient exponential decay dies down eventually, making the σ an insignificant component of s. It essentially states that the steady state Laplace Transform is the same as the Fourier Transform. Also in real life, if we are dealing with only frequency functions as input to a system, the exponent decay is insignificant and can be ignored for all practical purposes. The Transform is then reduced to a frequency response function only, where $s = j\omega$, as shown in the following equation:

$$\frac{V_{out}(j\omega)}{V_{in}(j\omega)} = H(j\omega)$$

Another conclusion we can draw is that the convolution process (where impulse response is convolved with the input function) is now converted into a multiplication operation: the Transform of the impulse response multiplied with the Transform of the input function produces the Transform of the output.

$$V_{out}(j\omega) = H(j\omega) \times V_{in}(j\omega)$$
$$v_{out}(t) = h(t) * v_{in}(t)$$

See Chapter 4, "Solutions of Differential Equations," for the description of the convolution process.

Filters, whether analog or digital, play a significant role in communication system design. Radio, telephony, and voice synthesis all have filters to either amplify or suppress a specific frequency or range of frequencies. A system with energy storage elements acts as a filter, and the solution of the governing differential equation is the filter's response. The Laplace Transform of the response (commonly known as frequency response) is a model for the filter and provides a method for analyzing the behavior of the system. We have devoted a complete chapter on filter design, but for now, we will present a simple low-pass filter as an example to show how the Transfer Function helps us analyze a system's response.

Filter Design and the Transfer Function

Consider the response of the simple RC circuit shown in Figure 5.6. Following in Equation 5.18 is the Transfer Function or the *impulse response* of the system where we have eliminated the exponential decaying factor from the variable $s = \sigma + j\omega$.

$$\frac{V_{out}}{V_{in}} = H(j\omega) = \frac{1}{1+j\omega RC} \tag{5.18}$$

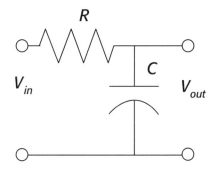

FIGURE 5.6 A simple low-pass filter with one resistor and one capacitor.

Equation 5.18 is a complex number, the real and imaginary components of which may be extracted by multiplying the numerator and denominator by $1 - j\omega RC$

$$H(j\omega) = \frac{V_{out}}{V_{in}} = \frac{1}{1+j\omega RC} \times \frac{1-j\omega RC}{1-j\omega RC} = \frac{1-j\omega RC}{1+(\omega RC)^2}$$

Thus, the real component is

$$\operatorname{Re}\{H(j\omega)\} = \frac{1}{1+(\omega RC)^2}$$

and the imaginary component is

$$\operatorname{Im}\{H(j\omega)\} = \frac{-j\omega RC}{1+(\omega RC)^2}$$

The magnitude and phase as a function of frequency may be defined as follows. The magnitude is shown in Equation 5.19.

$$|H(j\omega)| = \sqrt{\operatorname{Re}^2 + \operatorname{Im}^2} = \sqrt{\frac{1+(\omega RC)^2}{(1+(\omega RC)^2)^2}} = \frac{1}{\sqrt{1+(\omega RC)^2}} \quad (5.19)$$

And the phase:

$$\angle H(j\omega) = \tan^{-1}\left(\frac{\operatorname{Im}}{\operatorname{Re}}\right) = -\tan^{-1}\omega RC = \theta$$

For circuit components of $C = 1\mu f$, $R = 1k\Omega$, Equation 5.19 is reduced to Equation 5.20.

$$|H(j\omega)| = \frac{1}{\sqrt{1+\omega 10^{-3}}}, \theta = -\tan^{-1}(\omega 10^{-3}) \quad (5.20)$$

$$\angle\theta = -\tan^{-1}(\omega 10^{-3})$$

Figure 5.7 is a plot of the magnitude and phase response of Equation 5.20.

For a given input frequency $Vin = A\sin(\omega_0 t + \phi)$, the output is computed as,

$$Vout = |H(j\omega)|Vin = \frac{A}{\sqrt{1+(\omega_0 RC)^2}}\sin(\omega_0 + \phi - \tan^{-1}\omega_0 RC)$$

Let's analyze the result of Equation 5.20 (using different input frequencies) for the sake of comparison, and see if there could be any uniformity in the way the results are presented.

164 Digital Signal Processing Fundamentals

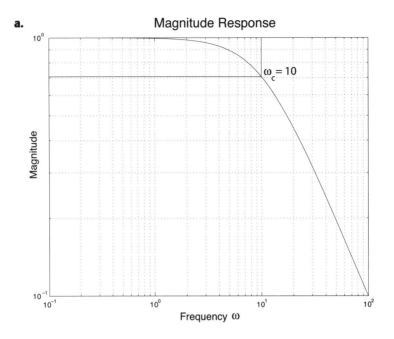

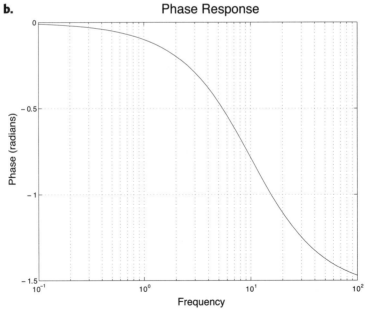

FIGURE 5.7 a) The magnitude response of the low-pass filter circuit of Figure 5.6. b) The phase response of the circuit.

Input frequency $\omega = 1 rad/sec$ is shown in Equation 5.21.

$$|H(j1)| = \frac{1}{\sqrt{1+(1\times 10^{-3})}} \approx 0.99 \qquad (5.21)$$

Input frequency $\omega = 10 rad/sec$ is shown in Equation 5.22.

$$|H(j10)| = \frac{1}{\sqrt{1+(10\times 10^{-3})}} \approx 0.99 \qquad (5.22)$$

Input frequency $\omega = 100 rad/sec$ is shown in Equation 5.23.

$$|H(j100)| = \frac{1}{\sqrt{1+(100\times 10^{-3})}} \approx 0.95 \qquad (5.23)$$

Input frequency $\omega = 1000 rad/sec$ is shown in Equation 5.24.

$$|H(j1000)| = \frac{1}{\sqrt{1+(1000\times 10^{-3})}} \approx 0.707 \qquad (5.24)$$

Input frequency $\omega = 10000 rad/sec$ is shown in Equation 5.25.

$$|H(j10000)| = \frac{1}{\sqrt{1+(10000\times 10^{-3})}} \approx 0.305 \qquad (5.25)$$

As you can see, there is hardly any effect on the magnitude when the input frequency is below $\omega = 1000 rad/sec$ (see Equations 5.21 through 5.25). But then as the frequency is increased, the gain is reduced considerably and at $\omega = 10000 rad/sec$ the magnitude is reduced to only .305 of the original input (see Equation 5.25). The circuit can easily be qualified as a low-pass filter that would not disturb input frequencies below $\omega = 1000 rad/sec$, but then it would start suppressing the higher frequencies above $\omega = 1000 rad/sec$. There is a clear trend in the output. Up until Equation 5.24, the magnitude remains unaffected (only dropping to 0.707), but then there is a steep drop beyond that point. You will see later in the section about poles and zeros that the mark of 0.707 is chosen as a turning point in the frequency response plot and is being used as a thumb rule in design practice.

So far, we have studied systems with very few energy storage elements. In an *RC* circuit there was only one (the capacitor), and in an *RLC* circuit there were barely two elements, the inductor and the capacitor. It was not too difficult to analyze them by the conventional means of algebra, but it was obvious that a more complex system would produce a more complex response. The following is a method to reduce a complex solution to a manageable algebraic form.

Reducing a Transfer Function's Complexity

In general, a Transfer Function of a more complex system can be described as a ratio of two polynomials,

$$H(s) = \frac{P(s)}{Q(s)} = \frac{z_0 s^m + z_1 s^{m-1} + .. z_{m-1} s + z_m}{p_0 s^n + p_1 s^{n-1} + .. p_{n-1} s + p_n}$$

Where $P(s)$ is a numerator polynomial in variable s and coefficient z, and $Q(s)$ is a denominator polynomial in variable s with coefficient p.

An alternate way of writing the polynomials in factored form is shown in Equation 5.26.

$$H(s) = \frac{P(s)}{Q(s)} = K \frac{(s-z_1) \times (s-z_2) \times .. \times (s-z_m)}{(s-p_1) \times (s-p_2) \times .. \times (s-p_n)} \qquad (5.26)$$

The numerator coefficients z_i are called zeros of the Transfer Function and p_i are called the poles of the Transfer Function, whereas the K is a real number scaling factor (Equation 5.26). The terminology *poles* and *zeros* is used for an apparent reason that at $s=z_i$ the numerator becomes zero, making the Transfer Function a zero, and at $s=p_i$ the denominator becomes zero, making the Transfer Function an infinite value, which causes the graph to look like a pole of a tent—if there was a graph of $H(s)$. But we are mainly interested in the magnitude of the Transfer Function, and since s is an imaginary number and p or z may be real or imaginary, so they have to be handled with the rules of complex number algebra. The poles and zeros have other significance, which will become apparent in the next section.

Poles, Zeros, and Steady State Frequency Response

When a Transfer Function is presented in the form of a numerator and denominator polynomial (as shown in Equation 5.26), each term in parentheses is a complex number vector, the z_1 and p_1 could be real or complex numbers, and the quantity s is equal to $\sigma + j\omega$. For a steady state frequency response, the quantity s may be replaced with $j\omega$ by ignoring the exponential decay σ, and the Equation 5.26 can be written as in Equation 5.27.

$$H(j\omega) = \frac{P(s)}{Q(s)} = K \frac{(j\omega - z_1) \times (j\omega - z_2) \times .. \times (j\omega - z_m)}{(j\omega - p_1) \times (j\omega - p_2) \times .. \times (j\omega - p_n)} \qquad (5.27)$$

The plot of the magnitude of the Transfer Function of Equation 5.27 versus the frequency describes a relationship between the output response and the input frequency. A more appropriate way of describing Equation 5.27 is,

$$|H(s)| = K \frac{(\sqrt{\omega^2 + z_1^2}) \times (\sqrt{\omega^2 + z_2^2}) \times .. \times (\sqrt{\omega^2 + z_i^2})}{(\sqrt{\omega^2 + p_1^2}) \times (\sqrt{\omega^2 + p_2^2}) \times .. \times (\sqrt{\omega^2 + p_i^2})}$$

The significance of poles and zeros is that at a specific frequency of $\omega = p_i$ (called pole frequency), the magnitude of the Transfer Function drops down to 0.707 of the magnitude at the start frequency of $\omega = 0$; similarly at frequency $\omega = z_i$ (called zero frequency), the Transfer Function magnitude is raised by 1.404 of the magnitude at the start frequency. The poles and zeros are only a convenient way of identifying the turning points in the magnitude of the Transfer Function, since at frequencies less than the poles' and zeros' frequencies, there is hardly any impact on the magnitude, but at frequencies above the poles and zeros, there is a rapid drop in the magnitude (in the case of poles) and rapid rise in magnitude in the case of zeros. (The poles and zeros are essentially vectors, whereas pole and zero frequencies are only indicative of the turning point in magnitude of the response.)

Let's take the simple case in which z_i and p_i are real (Equation 5.28). The Transfer Function of the low-pass RC filter of the previous example was already in the form of Equation 5.21 and had no zero, a pole at $j\omega = 10^3$, and the gain factor $K = 10^3$.

$$H(j\omega) = 10^3 \times \frac{1}{(j\omega + 1000)} \tag{5.28}$$

Calculating the magnitude of the Transfer Function at different frequencies,

$$\omega = 0, \omega = 100, \omega = 1000 \text{ and } \omega = 10000$$

$$H(j0) = 10^3 \times \frac{1}{\sqrt{0^2 + 1000^2}} = 1$$

$$H(j100) = 10^3 \times \frac{1}{\sqrt{100^2 + 1000^2}} = 0.995$$

$$H(j1000) = 10^3 \times \frac{1}{\sqrt{1000^2 + 1000^2}} = 0.707$$

$$H(j10000) = 10^3 \times \frac{1}{\sqrt{10000^2 + 1000^2}} = 0.1$$

Notice, at the pole frequency ($\omega = 1000$), the gain is 0.707 of its peak magnitude.

Figures 5.8a and b show the magnitude and the phase response of the Transfer Function of the pole vector of Equation 5.28.

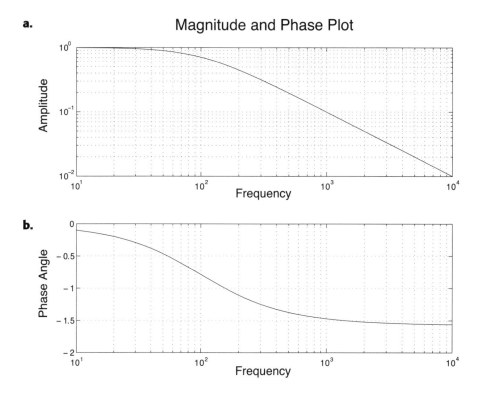

FIGURE 5.8 The magnitude and phase plot of the single pole Transfer Function of Equation 5.28 on a logarithmic scale.

It should be noted that when poles and zeros are complex numbers, their net effect on the magnitude is little distorted. The complex vectors always come in conjugate pairs. The tails lie not at the real axis but somewhere in the region $-\alpha + j\omega$ and $-\alpha - j\omega$ as shown in Figure 5.9b. Let's consider the parallel RLC circuit of Figure 5.9a; its Transfer Function is given in Equation 5.29.

$$H(s) = \frac{1}{C} \frac{s}{s^2 + 1/RC \, s + 1/LC} \tag{5.29}$$

Substituting $\alpha = 1/2RC$, $\omega_d = \sqrt{\omega_0^2 - \alpha^2}$, $\omega_0^2 = 1/LC$, we can rewrite Equation 5.29 in the factored form as in Equation 5.30.

$$H(s) = \frac{1}{C} \frac{s}{[s-(-\alpha+j\omega_d)][s-(-\alpha-j\omega_d)]} \quad (5.30)$$

Equation 5.30 has one zero at $s = 0$ and two poles at $s = (-\alpha + j\omega_d)$ and $s = (-\alpha - j\omega_d)$. With the more appropriate form of magnitude response, Equation 5.30 is simplified as

$$|H(j\omega)| = \frac{1}{C} \frac{j\omega - 0}{[j\omega-(-\alpha+j\omega_d)][j\omega-(-\alpha-j\omega_d)]}$$

The frequency response is shown in Figure 5.9c and the vector magnitudes $(s - (-\alpha + j\omega_d)$ and $s - (-\alpha - j\omega_d))$ are plotted in Figure 5.9b. The combined effect of the conjugate pair is such that the overall gain is more at the pole frequencies than at 0 frequency (enhancing a band of frequencies and suppressing the rest). It is obvious that we can select circuit components (*RLC*) such that at certain frequencies, the gain is much higher compared to the rest of the frequencies, a property that is exploited in the design of band-pass filters in the next chapter. Thus, a rule of thumb is devised that poles and zeros define turning points in the frequency response of a system.

We will discuss poles and zeros further in the next section, but for now let's talk about the range of numbers we are dealing with. The frequencies of interest can vary from as low as 0 to as high as megahertz. A linear scale certainly will not do the job. It is easier to describe the magnitude and phase response of the system using logarithmic scale, as it broadens the view of the response plot. Taking a logarithm of both sides of Equation 5.27, we get Equation 5.31.

$$|H(j\omega)| = |K| \times |j\omega - z_0| \times |j\omega - z_1| \times ... \times \frac{1}{|j\omega - p_0|} \times \frac{1}{|j\omega - p_1|}$$

$$\log|H(j\omega)| = \log|K| + \log|j\omega - z_0| + ... + \log\frac{1}{|j\omega - p_0|} + ... + \log\frac{1}{|j\omega - p_i|} \quad (5.31)$$

The logarithmic scale not only enhances the range of numbers that we can deal with but also converts the multiplication operation into an addition operation, simplifying the operation of computing the Transfer Function.

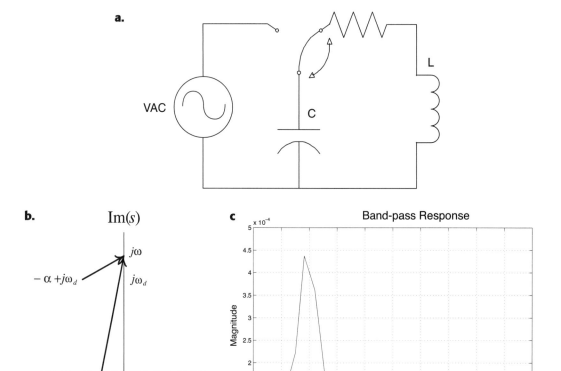

FIGURE 5.9 a) Series *RLC* circuit. b) The pole diagram of a complex conjugate pole vector. c) A plot of the frequency response of the *RLC* circuit.

Decibel dB

The Transfer Function is a *dimensionless* ratio, i.e., one of output and input, such as voltage or current. The magnitude is either less than unity, indicating a loss, or greater than unity, indicating gain. The decibel (abbreviated dB) is a special unit described as 10 times the log of the ratio of output power, delivered by the system and the power given to the system, as shown in Equation 5.32.

$$\text{Power gain} = 10\log\frac{P_{out}}{P_{in}} = 10\log\left[\frac{(v_{out}^2/R_{out})}{(v_{in}^2/R_{in})}\right] = 20\log\left|\frac{v_{out}}{v_{in}}\right| + 10\log\left|\frac{R_{out}}{R_{in}}\right| \quad (5.32)$$

Since, we always try to match the input and output impedance of the system, the last quantity in Equation 5.32 is considered 0, and the gain is usually described as

$$20\log\left|\frac{v_{out}}{v_{in}}\right|.$$

When we speak of Transfer Function in terms of *dB*, we mean 20 times the log of the magnitude of the Transfer Function. Thus, a gain of 10 is 20 dB, a gain of 100 is 40 dB, a gain of 1000 is 60 dB, a gain of 0.707 = −3dB, etc. With the help of decibel (as our unit of magnitude), we can describe quantities as small a number as 0.0001, all the way to 1,000,000, with only 10 divisions on a plot. You will see some examples shortly, but first we will discuss one last thing in terms of displaying the plot of the magnitude of the Transfer Function versus frequency: the Bode plot.

The Bode Plot

A bird's eye view of the system's frequency response can easily be calculated with a fair degree of approximation, if we consider what happens to the gain at the turning point of its poles and zeros. Consider the example of $H(j\omega) = 10^3 \times \frac{1}{j\omega - (-1000)}$ and the dB value of the gain and phase at various frequencies, shown in Table 5.2.

TABLE 5.2 Magnitude and Phase of the Transfer Function of the Expression $\frac{1}{j\omega - (-1000)}$

ω	$|H(j\omega)|$	$\angle H(j\omega)$	$20\log|1000|$	$20\log\left|\frac{1}{j\omega+1000}\right|$	$20\log|H(j\omega)|$
1 rad/sec	0.99	0 deg	60 dB	−60 dB	0 dB
10 rad/sec	0.99	0 deg	60 dB	−60 dB	0 dB
100 rad/sec	0.95	5.7 deg	60 dB	−60 dB	0 dB
1000 rad/sec	0.707	45 deg	60 dB	−63 dB	−3 dB
10000 rad/sec	0.1	85 deg	60 dB	−80 dB	−20 dB
100000 rad/sec	0.01	89 deg	60 dB	−100 dB	−40 dB
1000000 rad/sec	0.001	90 deg	60 dB	−1000 dB	−60 dB

The contribution of poles towards the gain of the Transfer Function (Column 5 of Table 5.2) is such that at frequencies less than the pole frequencies, the slope is approximately equal to 0 (basically a flat response), but at frequencies above the turning point (the pole frequency), the slope is approximately –20dB per decade. Similarly, for zeros, the slope is flat for frequencies less than the zero frequency, but anything beyond that the slope is +20dB per decade.

With this in mind, we can take individual poles and zeros and sketch out the independent response and linearly add them all later to obtain the final response. The steps are highlighted in Figure 5.10 for the following example, which shows a Transfer Function of two poles and two zeros,

$$H(s) = \frac{(s+100)(s+1000)}{(s+100)(s+8)}$$

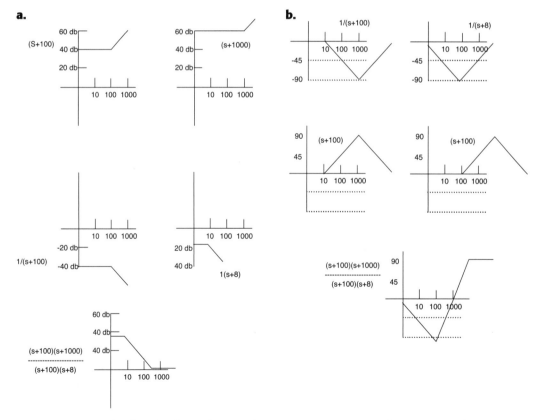

FIGURE 5.10 Bode plot of poles and zeros, drawn on approximate scale. a) Magnitude response. b) Phase response of poles and zeros.

Once a frequency response is identified that meets the requirements, the design job is complete. The poles and zeros are the design criteria as they specify the values of the circuit components you need to fulfill the design. There are standard circuits that fit each pole and zero, and you will see the examples later when we design analog filters. With this, we conclude the discussion of the Laplace Transform. Next, we discuss the digital counterpart of the Laplace Transform: the z-Transform.

THE Z-TRANSFORM

The z-Transform is the digital domain counterpart of the analog domain Laplace Transform. So far, all the mathematics of the solutions of differential equations revolved around the behavior of the physical components of resistors, capacitors, and inductors. But in real life, the problems are defined in terms of response, and the challenge is to design the circuits that match the response. Suppose the problem is to remove all the frequencies above 20 KHz from an incoming signal: the solution would be to pick the low pass filter circuit and simply choose the component values that match the Transfer Function of the circuit. Computers could do all that just as easily, if they were fast enough. If it weren't for the speed problem, we would probably never use the physical components in our systems. After all, every memory location of the computer could be used as an energy storage device, making it possible to create a system that is a thousand times more complex.

There is a correlation between the speed of execution and the input frequency of the digital world. Speed affects the sampling rate, and the sampling rate determines the maximum input frequency the digital system can process. Just to give you perspective, a 20 kHz sound wave (the best humans can hear) requires a sampling period of 25 microseconds. That means, within this time interval, the computer must take the A/D converter reading, multiply with the necessary coefficients (as if it were solving a differential equation), and produce the output to a D/A converter. Modern-day computers can handle this task easily, so the audio frequency is pretty much within the range of the digital domain. But, for a radio frequency of megahertz range, this kind of processing is out of the question (at least for now). If you suspect there is a high-frequency noise in your system that is above the sampling rate, you must add a hardware filter to eliminate the noise before taking any data.

You will see shortly that the z-Transform of a difference equation gives us an expression in frequency domain, similar to the Laplace Transform. The expression can be used to analyze the system's behavior in response to various inputs. Since, engineering design is not about solving the actual difference equation but finding the system's response to various inputs, the Transfer Function is the mathematical way to express such a response.

In the digital world, the differential equation of the analog domain becomes the difference equation of the discrete time. The time t of the continuous time corresponds to discrete time $k\Delta t$ (where k is the sample number and Δt is the data acquisition time per sample). The frequency in radians per sec ω_t is now normalized with $\omega_s = \omega_t \Delta t$ (radians per second times seconds per sample) and instead of integral over the time interval dt, we use summation over the interval radians per sample, and instead of integrating from 0 to infinity, the range is from 0 to the number of samples in two times the Nyquist frequency. With these new terminologies, we replace the Laplace Transform with the definition of the z-Transform shown in Equation 5.33.

$$\int_{0-}^{\infty} f(t)e^{-st}dt \rightarrow \sum_{k=0}^{\infty} f(k)e^{-j\omega_s k}$$

$$Z[f(k)] = F(z) = \sum_{k=0}^{\infty} f(k)z^{-k} \tag{5.33}$$

Where $z = e^{j\omega}$ and $f(k)$ is the kth sample value.

Similar to the Laplace Transform, the z-Transform converts a time dependent function into a frequency dependent function, but the frequency, instead of radians per second, is now in radians per sample, $\omega_s = \omega_t \Delta t$. Think of the z-Transform as a spectrum of all the frequencies present in the sample $f(k)$ up to the Nyquist frequency. In general, the z-Transform of the input $x(k)$ is defined as $X(z) = \sum_{k=0}^{\infty} x(k)z^{-k}$ and the output $y(k)$ as $Y(z) = \sum_{k=0}^{\infty} y(k)z^{-k}$.

Inverse z-Transforms

Technically, if the z-Transform is a spectrum of all the frequencies in a sample, then the inverse z-Transform should give you the sample with all the component frequencies in it. The integral in Equation 5.34 accomplishes the reconstitution of the original sample.

$$x(n) = \frac{1}{2\pi j} \oint X(z)z^{n-1}dz \tag{5.34}$$

It is not necessary to recover the original function using the inverse z-Transform; we could simply rearrange the Transform in the form of recognizable poles and zeros and use the table lookup method similar to the Laplace Transform.

Our main goal in solving difference equation through the z-Transform is to develop a relationship between the input and output in the form of a Transfer Function, essentially, defining an expression for the impulse response of the system.

The following are some useful z-Transforms.

Unit Impulse Functions

The impulse function is defined only at $k = 0$, as shown in Equation 5.35.

$$Z[\delta(k)] = \sum_{k=0}^{\infty} \delta(k) z^{-k} = \delta(0) = 1 \qquad (5.35)$$

Unit Step Functions

A closed form expression can be derived using the sum of a geometric sequence, as shown in Equation 5.36.

$$\sum_{k=0}^{\infty} c^k z^{-k} = \frac{1}{1 - cz^{-1}} \qquad (5.36)$$

Substituting for $c = 1$ in Equation 5.36, for a unit step function, we get

$$Z[u(k)] = \sum_{k=0}^{\infty} z^{-k} = \frac{1}{1 - z^{-1}}$$

Exponential Functions

Substituting for $c = a^k e^{jkb}$ in Equation 5.36, we get Equation 5.37.

$$Z[a^k e^{jbk}] = \sum_{k=0}^{\infty} a^k e^{jbk} z^{-k} = \frac{1}{1 - ae^{jb} z^{-1}} \qquad (5.37)$$

Sine Functions

Using exponential form of the sine function, we get Equation 5.38.

$$Z[\frac{1}{2} a^k (e^{jbk} - e^{-jbk})] = \sum_{k=0}^{\infty} a^k e^{jbk} z^{-k} = \frac{1}{1 - \frac{1}{2} a(e^{jbk} - e^{-jbk}) z^{-1}}$$

$$Z[\frac{1}{2} a^k (\sin b)] = \frac{(a \sin b) z^{-1}}{1 - 2az^{-1} + a^2 z^{-2}} \qquad (5.38)$$

Cosine Functions

Using the exponential form of the cosine function we get Equation 5.39.

$$Z[\frac{1}{2}a^k(e^{jbk\Delta t}+e^{-jbk\Delta t})] = \sum_{k=0}^{\infty} a^k e^{jbk\Delta t} z^{-k} = \frac{1}{1-\frac{1}{2}a(e^{jbk\Delta t}+e^{-jbk\Delta t})z^{-1}}$$

$$Z[\frac{1}{2}a^k(\sin b)] = \frac{(1-az^{-1}\cos b\Delta t)}{1-2az^{-1}\cos b\Delta t + a^2 z^{-2}} \quad (5.39)$$

Table 5.3 is a table of z-Transforms of some commonly used functions. For a complete listing see *http://dspcan.homestead.com/files/ztran/ztob.htm*.

TABLE 5.3 Commonly Used z-Transforms

	f(k)	X(z)
1	δ(k) (impulse function)	1
2	u(k) (unit step)	$\frac{1}{1-z^{-1}}$
3	c^k (exponential)	$\frac{1}{1-\alpha^{-1}}$
4	k (ramp)	$\frac{z^{-1}}{(1-z^{-1})^2}$
5	$a^k \cos(bk\Delta t)$ cosine function	$\frac{(1-az^{-1}\cos(b\Delta t))}{(1-2az^{-1}\cos(b\Delta t)+a^2 z^{-2})}$
6	$a^k \sin(bk\Delta t)$ sine function	$\frac{(a\sin b)z^{-1}}{1-2az^{-1}+a^2 z^{-2}}$

Adapted from *Digital and Kalman Filtering*, S.M. Bozic.

The Difference Rule

The most important property of the z-Transform is its Difference rule. Similar to the Laplace Transform, which changes a differential into a multiplication operation (see the derivative of Equation 5.10) where the first derivative is s times the Transform and the second derivative is s^2 times the Transform and so on, with the same context the z-Transform also changes the first difference into z^{-1} and second difference into z^{-2} and so on, as shown in the following derivative:

$$Z[y_k - y_{k-1}] = \sum_{k=0}^{\infty}(y_k z^{-k}) - \sum_{k=0}^{\infty}(y_k z^{-(k-1)})$$

The summation in the second term can be expressed as,

$$Z[y_{k-1}] = z^{-1}\sum_{k=0}^{\infty} y_k z^{-k}$$

$$Z[y_{k-2}] = z^{-2}\sum_{k=0}^{\infty} y_k z^{-k}$$

The Difference rule is also known as the shift property of the z-Transform, as multiplying with z^{-1} shifts the Transform to the previous input and multiplying with z^{-2} shifts to the previous to previous input, etc.

Next, we will see how the Difference rule helps us in defining an algebraic relationship of the output and input in the form of a Transfer Function.

The Transfer Function in z-Transforms

A general difference equation can be written in the following way:

$$y_0(k) + b_1 y(k-1) + \ldots b_m y(k-M) = a_0 x(k) + a_1 x(k-1) + \ldots + a_n x(k-N)$$

Applying the Difference rule to the equation produces a polynomial in the power of z, shown in Equation 5.40.

$$z^0 Y(z) + b_1 z^{-1} Y(z) + \ldots b_m z^{-m} Y(z) = z^0 a_0 X(z) + a_1 z^{-1} X(z) + \ldots + a_n z^{-N} X(z) \quad (5.40)$$

Equation 5.40 can be simplified as shown in Equation 5.41.

$$Y(z)\left(1 + \sum_{m=1}^{M} b_m z^{-m}\right) = X(z)\sum_{n=0}^{N} a_n z^{-n} \quad (5.41)$$

From Equation 5.41, we get a Transfer Function $H(z)$ showing a relationship of the output $Y(z)$ and the input $X(z)$ as shown in Equation 5.42.

$$H(z) = \frac{Y(z)}{X(z)} = \frac{\sum_{n=0}^{N} a_n z^{-n}}{1 + \sum_{m=0}^{M} b_m z^{-m}} \quad (5.42)$$

The expression in Equation 5.42 is a rational polynomial in variable z. The expression is suitable for plotting the magnitude H(z) as a function of sampling frequency (radians per sample). This is similar to the Transfer Function of continuous time as a rational polynomial in s. The designers of the system could use the plot to see, for any given input frequency, what the output magnitude would look like. A simple example is the process of averaging input values to eliminate the ripples in the input data. This is like implementing a low-pass filter. In this scheme, every new input x(k) is added to the previous input x(k–1) and divided by 2 to produce the output y(k). An iterative computer algorithm could be developed to produce the output, as shown in

$$y_k = \frac{1}{2}(x_k + x_{k-1})$$

Taking the Transform of both sides:

$$Y(z) = (\frac{1}{2}X(z) + \frac{1}{2}z^{-1}X(z))$$

To determine the frequency response, we obtain the Transfer Function as

$$H(z) = \frac{Y(z)}{X(z)} = \frac{1}{2} + \frac{1}{2}z^{-1}$$

Setting $z = e^{j\omega \Delta t}$,

$$H(z) = \frac{1}{2} + \frac{1}{2}e^{-j\omega \Delta t} \rightarrow \frac{1}{2} + \frac{1}{2}(\cos(\omega \Delta t) - j\sin(\omega \Delta t))$$

The magnitude of the Transfer Function is obtained from multiplying with its complex conjugate, as shown in Equation 5.43.

$$|H(z)| = \sqrt{\{0.5(1+\cos(\omega \Delta t)\}^2 - \{0.5 j(\sin(\omega \Delta t)\}^2} \quad (5.43)$$

Simplifying the term in Equation 5.43, using the identity $\sin^2(\omega t) + \cos^2(\omega t) = 1$, The Transfer Function magnitude is shown in Equation 5.44.

$$|H(z)| = \sqrt{0.5 + 0.5\cos(\omega \Delta t)} \quad (5.44)$$

And the phase change is shown in Equation 5.45.

$$\Phi_H(\omega \Delta t) = \arctan\left(\frac{\sin(\omega \Delta t)}{(1+\cos(\omega \Delta t))}\right) \tag{5.45}$$

The magnitude of Equation 5.44 is plotted in Figure 5.11, and you can see that the algorithm does suppress high frequencies very well. The example is given only to express the usefulness of the Transfer Function plot, albeit a very simple one. If the system is complex and involves several terms in the polynomials, such as the one in Equation 5.40, a better way to represent the Transfer Function is in the form of poles and zeros, just like the Transfer Function of the continuous time.

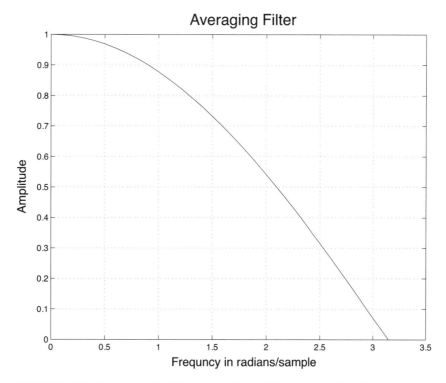

FIGURE 5.11 The averaging filter, gain plot at different sampling frequency.

Poles and Zeros of z-Transforms

A rational polynomial can be expressed as either a product or a sum of a partial fraction expansion, with real and imaginary coefficients. Presenting the polynomial as a product generates poles and zeros, similar to the continuous time Transfer Function. The terms in the poles and zeros correspond to vectors whose magnitude varies with the frequency with which the input values are multiplied. You will see an explanation of how the z-plane pole and zero magnitude varies as a function of frequency in Chapter 7, "Digital Filters," and we will make extensive use of them in our filter design process, but next is an example to prove the point that z-Transform is essentially a solution of difference equation.

Solving Difference Equations with z-Transform

Let's work through the example of the RC network as shown in Figure 5.12 for the component values of $R = 1$, $C = 1$, and $\Delta t = 0.1$ sec per samples and develop its discrete time impulse response using convolution and then the z-Transform.

Using x and y as the input and output, the difference equation is

$$RC\frac{dy}{dt} + y = x$$

$$RC\frac{y_k - y_{k-1}}{\Delta t} + y_k = x_k$$

$$y_k = \frac{y_{k-1}}{1 + \Delta t/RC} + \frac{\Delta t/RC}{1 + \Delta t/RC} x_k$$

Using the approximation $(1+\Delta t/RC)^{-1} \approx 1 - \Delta t/RC$ and neglecting the higher order terms, we get Equation 5.46.

$$y_k = a_0 x_k + b_1 y(k-1) \qquad (5.46)$$

Where $a_0 = \Delta t/RC, b_1 = (1 - a_0)$

The impulse response of the equation can be obtained by injecting a sample of unit impulse, as shown in Equation 5.47.

$$x(k) = \delta(k) \begin{cases} 1 \ldots k = 0 \\ 0 \ldots k \neq 0 \end{cases}$$

$$y(0) = h(0) = a_0$$
$$y(1) = h(1) = a_0 b_1$$
$$y(2) = h(2) = a_0 b_1^2$$
$$\vdots$$
$$y(k) = h(k) = a_0 b_1^k$$

$$y(k) = a_0 x(k) + a_0 b_1 x(k-1) + a_0 b_1^2 x(k-2) + \ldots + a_0 b_1^k x(0) \tag{5.47}$$

The same result can be obtained through expanding the z-Transform of Equation 5.46 as shown in Equation 5.48.

$$y_k = a_0 x_k + b_1 y(k-1)$$

$$Y(z) = a_0 X(z) + b_1 z^{-1} Y(z)$$

$$H(z) = \frac{Y(z)}{X(z)} = \frac{a_0}{1 - b_1 z^{-1}} \tag{5.48}$$

We can use the analogy of a geometric sequence, whose expansion is given as

$$\frac{1}{1 - cz^{-1}} = 1 + c^1 z^{-1} + + c^2 z^{-2} + \cdots + c^n z^{-n}$$

The result of Equation 5.48 may be expanded as in Equation 5.49.

$$Y(z) = a_0 X(z) + a_0 b_1^1 z^{-1} X(z) + a_0 b_1^2 z^{-2} X(z) \cdots a_0 b_1^k z^{-k} X(z) \tag{5.49}$$

Taking the inverse of Equation 5.49, we get Equation 5.50.

$$y(k) = a_0 x(k) + a_0 b_1 x(k-1) + a_0 b_1^2 x(k-2) + \ldots + a_0 b_1^k x(0) \tag{5.50}$$

The result of Equation 5.50 matches Equation 5.47. Both of them are solutions of the difference equation derived in Equation 5.46.

Our main purpose of going through the rigor of mathematics to find the solutions of differential equations is that they form the basis for realizing our main objective: designing the analog and digital filters, which will be discussed in the next chapter.

SUMMARY

The goal in this chapter was to derive a closed form expression of a solution of differential equations called a *Transfer Function*, which describes the relationship between the input frequencies and output response. The poles and zeros extracted from the Transfer Function describe a system's response to varying input frequencies, thus helping establish a relationship between the inputs and outputs of a linear system.

6 Filter Design

In This Chapter

- Filter Terminologies
- Filter Design Methods
- Analog Filters

A *signal* is any useful information and *noise* is the not so useful information, invariably, a part of the signal. The signals, seemingly arbitrary in nature, may be thought of as sinusoidal functions (the fundamental and the harmonics), as described by Fourier. You can imagine what would happen to signal frequencies, if they were passed through a system of energy storage elements (such as the linear time invariant systems we have been studying, which have within them an impulse response in the form of sinusoidal functions), obviously, the matching frequencies in the signal and the system would be enhanced, and the unmatched frequencies would subside. The result is a filter, separating the desired frequencies from the undesired. We can tailor the response of such systems to our wishes, by physically choosing the components (in case of an analog system) or the coefficients (in case of a digital system). This is the basics of the filter design: suppressing or enhancing a specific frequency or a set of frequencies by passing a signal source through a system of energy storage elements. Filters modify signals in ways specified by the filter's frequency response.

In essence, a filter is a system capable of generating response based on the frequency selection. An analog filter is an electrical network comprising physical components of the resistors, capacitors, and inductors, whereas, a digital filter is a computer software algorithm producing the filter response. Filters can either amplify or suppress a specific frequency or range of frequencies, but they can also be designed to create a pattern of frequencies as a response to a stimulus. The current trend is toward digital filter design, but still, analog filters are much in demand in the high-frequency arena as well as in feedback and control systems in which response time is critical.

We have discussed filters in the previous chapter on solutions of differential equations and have designed some simple low-pass and high-pass filters, using both the analog and digital techniques. The reason for a tandem discussion is that most analog filters are a precursor of digital filters and have roots behind the filter design mathematics. The criteria of any filter design are to establish the boundary of the desirable frequencies and then define how much suppression of the undesired frequencies is acceptable. Remember, frequencies can never be eliminated, they can only be suppressed. The first criterion provides the cutoff frequency, and the second one establishes how many poles and zeros are required in the filter design. The design methods are well established in both the analog and the digital domain, and the design procedures are a simple matter of selecting the appropriate components or coefficients. In this chapter, some commonly used designs of analog filters are discussed that become the prototype for our next chapter on digital filter design.

We begin our discussion with the explanation of terms used in both the analog and the digital filters.

FILTER TERMINOLOGIES

The following terminologies explain different aspects of filter characteristics. See Figure 6.1a and 6.1b for a reference.

Pass band: Frequency range preserved in the output
Stop band: Frequency range suppressed in the output

Gain: Amount of maximum amplification in the output

Transition band: The region of frequencies between pass band and stop band

Stop-band attenuation: The difference in dB between the pass-band and the stop-band gain

Pass-band ripple: The maximum fluctuation in the frequency response in the pass band

Stop-band ripple: The maximum fluctuation in the frequency response in the stop band

Roll-off rate: The steepness of the slope in the transition band (multiples of 20db/decade)

Order: The number of poles in the system function $H(s)$; the higher the order, the steeper the roll off

Cutoff frequency: The edge of the pass band (3db point)

Q: The sharpness of the peek in a band-pass filter

Filters are specified by the magnitude at two critical frequencies: the cutoff frequency ω_c, which signifies the end of the pass-band region, and the rejection frequency ω_r, which signifies the beginning of the stop-band region, and in between there is a transition band as shown in Figure 6.1a. The two magnitudes are specified as,

$$1 \geq |H(j\omega)|^2 > \frac{1}{1+\varepsilon^2} \qquad for\ |\omega| < \omega_c$$

$$|H(j\omega)|^2 < \frac{1}{1+A^2} \qquad for\ |\omega| > \omega_r$$

For example, a −3 db magnitude $|H(j\omega)|^2 = 0.5$ at the cutoff frequency indicates $\varepsilon = 1$ and the stop-band attenuation of −40 db $|H(j\omega)|^2 = .01$ determines $A = 9.9$. The filter order or the number of poles in the filter Transfer Function is determined by the quantities A and ε as shown later in the section that addresses Butterworth filter design.

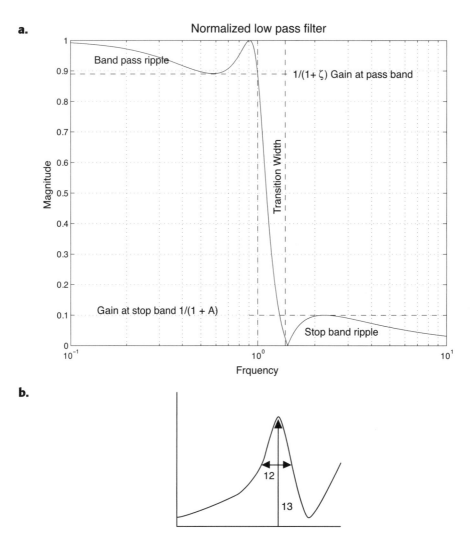

FIGURE 6.1 a) Different characteristics of a filter response. b) The cutoff frequency and Q of a band-pass filter.

FILTER DESIGN METHODS

We have seen in the previous chapter how a first and second order differential equation's response acts as a frequency selector. The filter design is essentially implementing solutions of differential equations, either through the Laplace Transform or the convolution process, but the modeling is done through the Transfer Function that formed the input and output relationship.

The higher frequencies are suppressed from the voltage across the capacitor in a network of resistors and capacitors. Similarly, the lower frequencies are suppressed from the voltage across the resistor in the network of resistors and capacitors, and a band-limited response is achieved with a second order section.

The concept of poles and zeros is central to the filter design, as they can be directly translated to the Transfer Function of the system. You have seen in the previous chapter, how magnitude of pole and zero vectors define the cutoff regions. The frequencies beyond the pole frequencies are reduced substantially, and the frequencies beyond the zero vectors are greatly enhanced. Being a linear system, the overall filter response is the sum of the individual poles and zeros response. Although any form and shape of the frequency response could be obtained by carefully selecting these vectors, the response generally falls into the following four basic types: low-pass, high-pass, band-pass, and band-stop filters. There are variations of filters such as notch filters, narrow band-pass, etc. that are special implementations of the band-pass and band-stop filters.

As the names suggest, the *low-pass* filters allow low frequency to pass through but stop high frequencies, the *high-pass* filters allow high frequencies to pass through but stop low frequencies, the *band-pass* filters allow a specific band of frequencies to pass through but stop all others, and the *band-stop* filters allow all frequencies to pass through except a certain band.

The fact that the LTI systems only alter the amplitude or phase but not the frequency itself, leaves us with only one choice and that is to suppress the undesired frequencies. Ideally, we would like to have the desired frequencies untouched while the undesired frequencies are eliminated, as shown in Figure 6.2, but this is not practical, as it requires infinite filter blocks, so we will leave the quest for achieving the ideal response only as a goal.

The filter design procedure has been standardized around a set of low-pass filters having the normalized cutoff frequency, $S_N = 1\,rad/sec$. From this prototype, low-pass filters of all other types may be derived with a simple substitution scheme, as shown in the following section.

Frequency-frequency Mapping

Here is a low-pass to low-pass transform (ω_c as the cutoff frequency):

$$s_N = s/\omega_c$$

Here is a low-pass to high-pass transform (ω_c as the cutoff frequency):

$$s_N = \omega_c/s$$

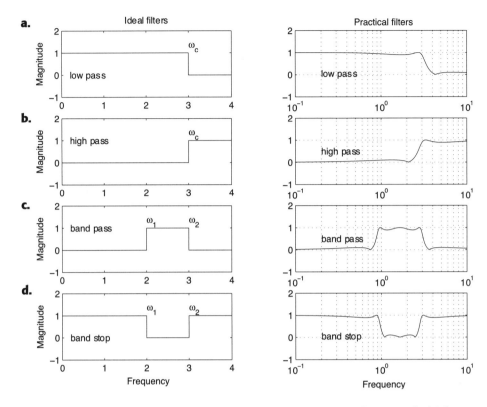

FIGURE 6.2 Ideal filter response and practical filter response: a) low pass, b) high pass, c) band pass, d) band stop.

Here is a low-pass to band-pass transform (bandwidth ω_1, ω_2):

$$s_N = (s^2 + (\omega_2\omega_1))/(s + (\omega_2 - \omega_1))$$

Here is a low-pass to band-stop transform (bandwidth ω_1, ω_2):

$$s_N = (s + (\omega_2 - \omega_1))/(s^2 + (\omega_2\omega_1))$$

For example, a normalized low-pass filter whose Transfer Function is given as $H(s) = 1/(1 + s_N)$ may be converted into another low-pass filter with the cutoff frequency $\omega_c = 50 rad/sec$ by substituting S_N with S/ω_c. The resulting filter Transfer Function is $H(s) = 50/(50 + s)$. Similarly, a second order normalized low-pass filter Transfer Function such as $H(s) = 1/(s^2 + \sqrt{2}s + 1)$ may be converted into a band-stop filter with a bandwidth of 10 rad/sec centered around 50 rad/sec, using the low-pass to band-stop mapping as

$$H(s) = \frac{\left(5760000 + 4800 s^2 + s^4\right)}{\left(5828282 + 3434 s + 4829 s^2 + 1.4142136 s^3 + s^4\right)}$$

All these calculations could be performed easily with the help of Scilab instructions, as follows (see Appendix B, "Scilab Tutorial," for a tutorial on Scilab).

```
‡ s=poly(0,'s');
‡nlp=1/(1+sqrt(2)*s+s^2);
‡bsmap=(s+(60-40))/(s^2+(60*40));
‡horner(lp,lpmap)
```

The Scilab function horner performs the required transformation.

Similarly, the Matlab script fig6_1a.m (provided in the accompanying CD-ROM directory /matlab/fig6_1a.m) performs the desired transformation, as shown in Figure 6.1a.

The digital and analog filters may be identical in their response up to certain frequencies, but the implementation of each is an entirely different matter. For analog filters, the Transfer Function is implemented as a circuit design with the coefficient values translated to the component values of the resistors, capacitors, and inductors (op-amps may be added in place of inductors and to isolate the blocks of circuits from one another). On the other hand, the digital filters are realized by transforming the Transfer Function into a difference equation that is to be solved with convolution using an iterative algorithm. There is no limit to the range of input frequencies for analog filters, but the digital filters are limited in their frequency response: they can only process frequencies up to the Nyquist limit (which is defined as half the rate of the sampling frequency).

Every filter has a corresponding Transfer Function, but it is easier to implement filters in terms of the building blocks of Transfer Functions. It not only reduces the complexity of the overall system but gives a modular approach to the implementation process.

Building Blocks of Transfer Functions

We have seen in the previous chapter that each pole contributes to a 20 dB drop in the roll-off rate of the frequency response. The response is improved as we add more poles to the Transfer Function of the system, but then the order of the polynomial grows as we add more poles, making the system more complex to design.

One way to simplify the complexity is to cascade blocks of simple Transfer Functions. These blocks are made up of the first and second order systems of single real or single complex poles and zeros. Complex filters are implemented as cascades of the basic building blocks.

The following Transfer Functions provide the four basic types of filters, namely, the low-pass, high-pass, band-pass, and band-stop filters. The equivalent circuit matching the Transfer Function is described in the section about analog filters, and the equivalent difference equations for the digital filters are discussed in the next chapter.

First Order Low-Pass Filter Transfer Function

The Transfer Function of Equation 6.1 is a low-pass single pole filter with the cut-off frequency of p_1 and the gain of K at 0 frequency.

$$\frac{V_O}{V_I} = \frac{K}{1 + s/p_1} \tag{6.1}$$

The magnitude response is

$$|H(j\omega)| = \left|\frac{K}{1 + j(\omega/p_1)}\right| = \frac{K}{\sqrt{1 + (\omega/p_1)^2}}$$

The phase function from the Equation 6.1 is

$$\phi = -\tan^{-1}\frac{\omega}{p_1}$$

The frequency and the phase response of the low-pass Transfer Function (for $p_1 = 50$) are shown in Figure 6.3.

Second Order Low-Pass Filter Transfer Function

A 40 dB roll-off rate may be achieved with the addition of an extra pole resulting in a second order section as shown in the Transfer Function of Equation 6.2,

$$H(s) = \frac{k\omega_C^2}{s^2 + (\omega_C/Q)s + \omega_C^2} \tag{6.2}$$

NOTE

When $\omega_C/Q = \sqrt{2}$ in Equation 6.2 becomes a second order Butterworth polynomial, discussed later in the section, the normalized frequency response of the second order section compared with the first order is shown in Figure 6.4. Notice the improvement in roll-off rate of the second order Transfer Function.

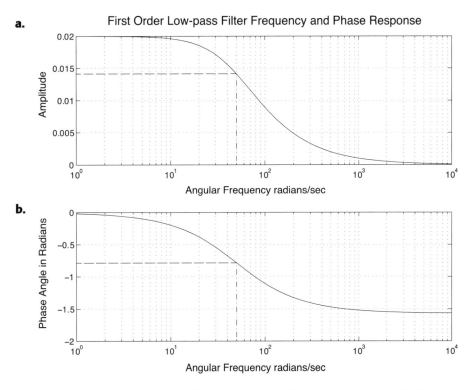

FIGURE 6.3 First order low-pass filter response. a) Magnitude response. b) Phase response.

First Order High-Pass Filter Transfer Function

The Transfer Function in Equation 6.3 is for a high-pass single-pole filter with the cutoff frequency of p_1 and the gain of K at 0 frequency.

$$H(s) = \frac{V_O}{V_I} = \frac{K}{1 + 1/sp_1} \quad (6.3)$$

The magnitude response is

$$|H(j\omega)| = \left| \frac{K}{1 - j(1/\omega p_1)} \right| = \frac{K}{\sqrt{1 + (1/\omega p_1)}}$$

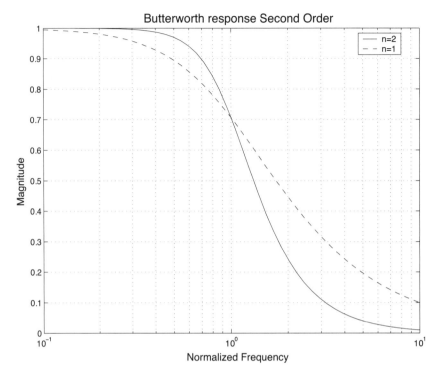

FIGURE 6.4 Second order low-pass filter magnitude response compared to the first order filter response.

The phase function from Equation 6.3 is

$$\phi = \tan^{-1} \frac{1}{\omega p_1}$$

The magnitude response is shown in Figure 6.5.

The Second Order High-Pass Filter Transfer Function

The second order gain of a high-pass filter is given as shown in Equation 6.4.

$$H(s) = \frac{V_O}{V_I} = \frac{K_3 s^2}{s^2 + b_1 s + b_0} \quad (6.4)$$

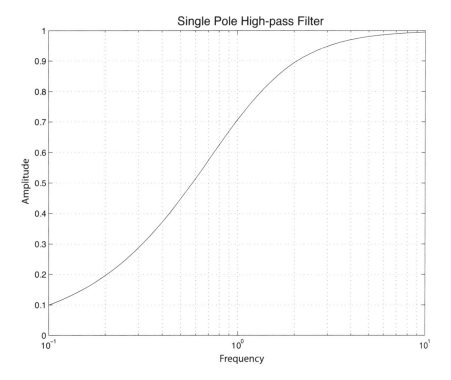

FIGURE 6.5 First order high-pass filter magnitude response.

The magnitude response is

$$|H(j\omega)| = \left| \frac{-K_3\omega^2}{(b_0 - \omega^2) + (j\omega b_1)} \right| = \frac{K_3\omega^2}{\sqrt{(b_0 - \omega^2)^2 + (\omega b_1)^2}}$$

The phase function from Equation 6.4 is

$$\tan^{-1} \frac{b_1\omega}{(b_0 - \omega^2)}$$

The normalized frequency and phase response of the second order high-pass transfer function is shown in Figure 6.6.

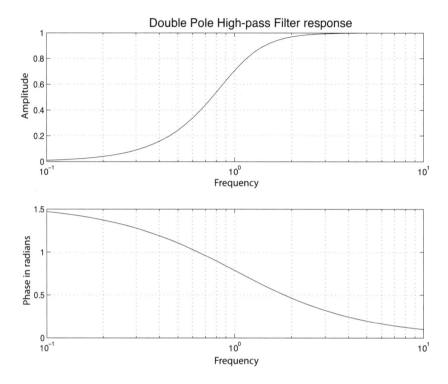

FIGURE 6.6 Second order high-pass filter response, the magnitude, and the phase response.

The Second Order Band-Pass Filter Transfer Function

A band-pass filter may be obtained from a low-pass filter Transfer Function with a simple substitution of S_N with $(s^2 + (\omega_2 \omega_1))/(s + (\omega_2 - \omega_1))$ as shown in Figure 6.1a. The band-pass filter may also be defined in terms of the center frequency ω_0 and its two side frequencies $\omega_{3\,dB}$ where the gain falls off to the −3 dB of the center frequency. The Transfer Function is given as shown in Equation 6.5.

$$H(s) = \frac{V_O}{V_I} = \frac{K_3 s}{s^2 + (\omega_0/Q)s + \omega_0^2} \tag{6.5}$$

Where $Q = (\omega_0/\omega_{3db})$.

The width of the bandwidth is controlled by the quantity Q. If a narrow bandwidth is desirable, set the Q to a high value such as 10 or more.

The magnitude response is

$$|H(j\omega)| = \left|\frac{K_3\omega}{(\omega_0^2 - \omega^2) + j(\omega_0/Q)}\right| = \frac{K_3\omega}{\sqrt{(\omega_0^2 - \omega^2)^2 + j(\omega_0/Q)^2}}$$

The phase function from Equation 6.5 is

$$\tan^{-1}(j\omega) - \tan^{-1}\frac{j(\omega_0/Q)}{(\omega_0^2 - \omega^2)}$$

The magnitude response is shown in Figure 6.7.

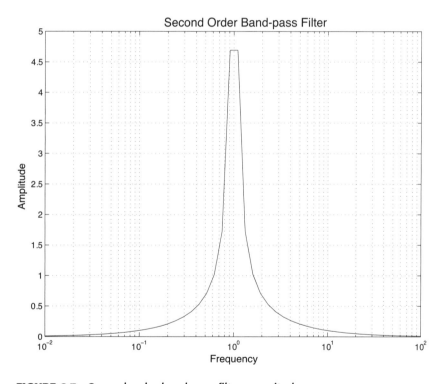

FIGURE 6.7 Second order band-pass filter magnitude response.

The Second Order Band-Stop Filter Transfer Function

The band-stop filter is also known as the notch filter. Similar to the band-pass filter, the Transfer Function is defined in terms of the center frequency ω_c that would be removed from the system and the two side frequencies ω_0 with –3 dB gain. The Transfer Function is given as shown in Equation 6.6.

$$H(s) = \frac{V_O}{V_I} = \frac{K_4(s^2 + \omega_c)}{s^2 + (\omega_0/Q)s + \omega_0^2} \qquad (6.6)$$

Where $Q = (\omega_0/\omega_{3db})$.

The magnitude response is

$$|H(j\omega)| = \left| \frac{K_4\sqrt{(\omega_c^2 + \omega^2)^2}}{(\omega_0^2 - \omega^2) + j(\omega_0/Q)} \right| = \frac{K_4(\omega_c^2 + \omega^2)}{\sqrt{(\omega_0^2 - \omega^2)^2 + j(\omega_0/Q)^2}}$$

The phase function from the Equation 6.6 is

$$\tan^{-1}(\omega_c^2 + \omega^2) - \tan^{-1}\frac{j(\omega_0/Q)}{(\omega_0^2 - \omega^2)}$$

The magnitude response is shown in Figure 6.8.

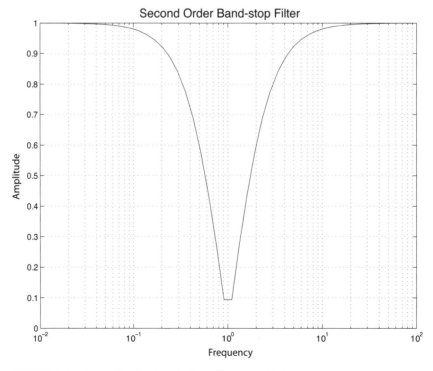

FIGURE 6.8 Second order band-stop filter magnitude response.

Systems often require more sophisticated filter response than the 3 dB gain and 20 dB per decade drop in the roll-off rate as provided by the first order Transfer Function polynomials. There are different methods that enhance different aspects of the filter characteristics. The Chebyshev filters improve upon the roll-off rate but add ripples in the pass-band region, whereas the Butterworth filters give flatter response in the pass-band region but are not very efficient in the roll-off rate. We will discuss the Butterworth filters in the next section and the Chebyshev filter in the coming chapter.

Butterworth Filters

The goal in any filter design is to get as close as possible to the ideal filter response. We would like to see a flatter response in the pass-band region and a steeper roll-off in the transition band. The Butterworth polynomials offer such a response. The response function is given as

$$H(s) = \frac{1}{1+s^n}$$

And the magnitude squared is shown in Equation 6.7.

$$|H(j\omega)|^2 = \frac{1}{1+(j\omega)^{2n}} \quad (6.7)$$

Notice the coefficient of $j\omega$ has been chosen as 1, this is equivalent of normalized cutoff frequency of $\omega_o = \frac{\omega}{\omega_{Cutoff}} = 1$. Having the normalized response simplifies the design process by designing the circuit for unit response, and once all the components are identified, in the last step the values are scaled for the desired response. At cutoff frequency when $\omega = \omega_{Cutoff}$, the magnitude response of Equation 6.7 becomes 0.707 for all values of n. The contribution becomes less significant for $\omega_o < 1$, and for $\omega_o > 1$, the magnitude approaches 0 faster with increasing n. The result is a flatter response for frequencies less than the cutoff frequency and steeper roll for the frequencies greater than the cutoff frequency with increasing n, as shown in the Figure 6.9. Through the normalized low-pass filter (as defined in Equation 6.7), all other types of filters may be derived with the help of the frequency-frequency mapping mentioned earlier.

The Butterworth expansion for $N = 1$ is given as

$$|H(j\omega)|^2 = \frac{1}{1+(\omega/\omega_c)^2} \quad (6.8)$$

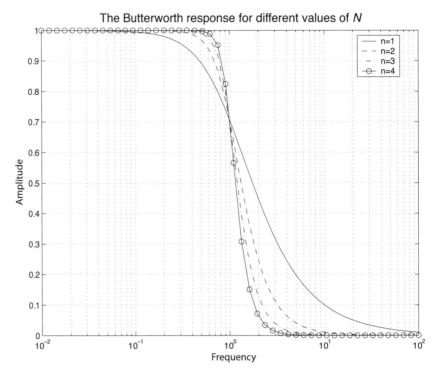

FIGURE 6.9 The Butterworth response with different values of n.

Where ω_c is the cutoff frequency, as $\omega \to \infty$ the response approaches 0, but when ω gets closer to 0, the response becomes 1. In between, there is a gradual loss of magnitude. Precisely at the cutoff frequency of ω_c, the magnitude is −3 dB. This is equivalent to the first order Butterworth response as shown in Figure 6.4. The expansion of coefficients $(\omega)^{2N}$ may be explained as follows.

Butterworth Expansion

If the power vectors of the pole frequency $(\omega_C)^{2N}$ were to be drawn on a circle (with the radius (ω_C)) that can be normalized and set equal to 1, all vectors would appear as equally spaced on a semicircle. Since multiplying a unit vector to itself simply shifts the vector to an angle, the roots of Equation 6.7 are all shifted by an angle π/N with respect to the real axis. This could be evident from the following analogy.

The poles are actually N roots of the denominator polynomial of the Transfer Function, as shown in Equation 6.9.

$$H(s) = \frac{1}{1+s^N}$$

$$1 + s^N = 0$$

$$s^N = -1, s = \sqrt[N]{e^{j\pi}} = \sqrt[N]{-1}$$

$$s = \cos(\pi) + j\sin(\pi)$$

$$H(j\omega)H(-j\omega) = |H(j\omega)|^2 = 1 + (-1)^n s^{2N} = 0 \quad (6.9)$$

Where s is a complex value $s = e^{j\pi}$ evaluated on a unit circle. See Chapter 2 for an explanation of the root of complex numbers.

Figure 6.10 indicates pole locations for various values of N. Only the poles on the left side of the circle are taken as the roots of the Transfer Function polynomial. Except for the first pole, all others are complex conjugates. If N is even, all poles are complex conjugates and when N is odd, there is only one real and the rest are complex conjugate poles. An nth order Butterworth Filter is defined as

$$|H(s)|^2 = \frac{1}{\prod_{i=1}^{N}(s-s_i)} = \frac{1}{(s-s_1)(s-s_2)\cdots(s-s_N)}$$

Where $s_i = e^{j\pi[2i+n-1/2n]} = \cos(\pi\frac{2i+n-1}{2n}) + j\sin(\pi\frac{2i+n-1}{2n})$

The following calculations indicate the pole vectors for different values of N:

$N = 1$
$$s_1 = \cos(\pi) + j\sin(\pi) = 1$$

$$H(s) = \frac{1}{(s+1)}$$

$N = 2$
$$s_1 = \cos(\pi 3/4) + j\sin(\pi 3/4) = -.707 + j.707$$

$$s_2 = \cos(\pi 5/4) + j\sin(\pi 5/4) = -.707 - j.707$$

$$H(s) = \frac{1}{(s_2^2 + \sqrt{2}s_2 + 1)}$$

$N = 3$

$s_1 = \cos(\pi 2/3) + j\sin(\pi 2/3) = -.5 + j.866$

$s_2 = \cos(\pi) + j\sin(\pi) = 1$

$s_3 = \cos(\pi 2/3) + j\sin(\pi 2/3) = -.5 - j.866$

$$H(s) = \frac{1}{(s_3^3 + 2s_3^2 + s_3 + 1)} = \frac{1}{(s_3 + 1)(2s_3^2 + s_3 + 1)}$$

$N = 4$

$s_1 = \cos(\pi 5/8) + j\sin(\pi 5/8) = -.382 + j.924$

$s_2 = \cos(\pi 7/8) + j\sin(\pi 7/8) = -.924 + j.383$

$s_3 = \cos(\pi 9/8) + j\sin(\pi 9/8) = -.924 - j.383$

$s_4 = \cos(\pi 11/8) + j\sin(\pi 11/8) = -.382 - j.924$

$$H(s) = \frac{1}{(s_4^4 + 2.61s_4^3 + 2.61s_4^2 + s_4 + 1)} = \frac{1}{(s_4^2 + 0.765s_4 + 1)(s_4^2 + 0.765s_4 + 1)}$$

The design parameter N or the filter order is calculated by the specifications of the magnitude at the cutoff frequency and the rejection frequency. For example, if the magnitude at the 20Hz is specified as –3 db and the attenuation at 30 Hz is –40 db, the filter order N is

$$-3db \rightarrow -\frac{3}{20} \rightarrow 10^{-\frac{3}{20}} \rightarrow \left|10^{-\frac{3}{20}}\right|^2 = \frac{1}{1+\varepsilon^2} = 0.5, \varepsilon = 0.707 = \frac{1}{1+[\omega_P/\omega_C]^{2N}}$$

$$-40db \rightarrow -\frac{40}{20} \rightarrow 10^{-\frac{40}{20}} \rightarrow \left|10^{-\frac{40}{20}}\right|^2 = \frac{1}{1+A^2} = 0.0001, A = 0.01 = \frac{1}{1+[\omega_S/\omega_C]}$$

$$N \geq \frac{\log(A/\varepsilon)}{\log(\omega_P/\omega_S)} \geq 10.5, N = 11$$

Next, we discuss the implementation details of the Transfer Functions in terms of the electronic circuit designs for the analog filters (the digital filters are discussed in the next chapter).

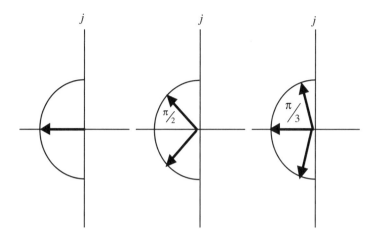

FIGURE 6.10 The pole locations of the Butterworth expansion.

ANALOG FILTERS

In an analog domain, a filter is a circuit that produces the expected response specified by the corresponding filter Transfer Function. We have seen earlier how a first and second order differential equation's response acts as a frequency selector. The higher frequencies are suppressed from the voltage across the capacitor in a network of resistors and capacitors. Similarly, the lower frequencies are suppressed from the voltage across the resistor in the network of a resistor and capacitor.

The filter circuitry in our designs will have basically two types of components, the passive components (the resistors and capacitors) and the active components (the op-amps), hence the name *active RC filters*. The op-amps play dual roles in the analog filters. They not only simulate inductors (if needed as a second order section) but also act as noninteracting blocks to prevent loading effects on the passive components. Since op-amps play an important role in setting up the network equations, a brief overview is presented next as a refresher.

Op-amps as Differential Amplifiers

A differential amplifier's function is to amplify the difference between two signals. The basic schematic of the op-amp is represented in Figure 6.11a, and the corresponding circuit model is shown in Figure 6.11b. The R_i is the differential input impedance of an extremely high value with practically no current flowing through it. The R_s is the output impedance, a negligible quantity as if it does not exist in the circuit. The amplifier gain of Figure 6.11a is expressed as

$$v_o = a_o(v_1 - v_2)$$

Where a_o is the open loop gain of the op-amp when there is no feedback and is usually very large in the range of 10^5 to 10^6, considered infinite for an ideal op-amp. We will use the equivalent of voltage controlled voltage source, as shown in Figure 6.11b, in the loop analysis of the network when we use op-amps in our filter design.

Two things need to be remembered when dealing with op-amps:

- Infinite input resistance means the current into the inverting input is zero:

$$i_- = 0$$

- Infinite gain means the difference between $v+$ and $v-$ is zero:

$$v_+ - v_- = 0$$

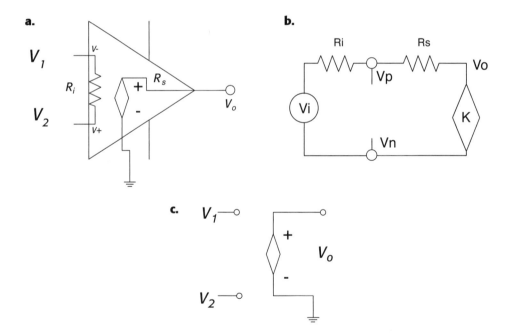

FIGURE 6.11 a) Op-amp as a differential amplifier. b) Equivalent circuit model. c) Input and output voltage across the op-amp.

The two most common configurations are provided next to help identify the loop equation created by a filter circuit.

Inverting Amplifier

Figure 6.12a shows an inverting amplifier configuration and Figure 6.12b shows the summation of current at the inverting input.

The current passing through the two resistors is equal as no current should be going through the op-amp,

$$i_1 = i_2 = i$$

Since all the current is passing through the source resistor R_1 and the feedback resistor R_2, we have the following voltage drop:

$$v_i = R_1 i \text{ and } v_0 = -R_2 i$$

Resulting in the amplifier voltage gain of

$$\frac{v_o}{v_i} = -\frac{R_2}{R_1}$$

The op-amp will provide whatever output voltage is necessary so that both the input voltages are equal.

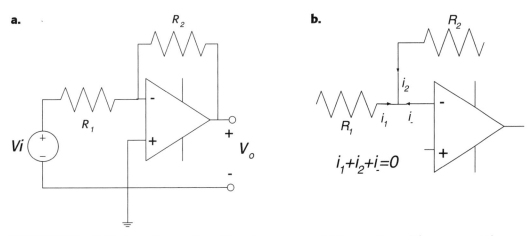

FIGURE 6.12 a) The inverting configuration of an op-amp. b) Summation of the current at the node.

Noninverting Amplifier

Figure 6.13 is a noninverting amplifier configuration showing the summation of the current at the non-inverted input.

Since no current is passing through any of the op-amp input, the two input voltages are equal:

$$v_+ = v_-$$

The voltage across $v_- = \frac{R_1}{R_1 + R_2} v_o$ and $v_+ = v_- = v_i$.

The voltage gain is

$$\frac{v_o}{v_i} = \frac{R_1 + R_2}{R_1} = 1 + \frac{R_2}{R_1}$$

Resulting in the voltage gain of

$$v_o = \frac{R_1 + R_2}{R_1} v_i$$

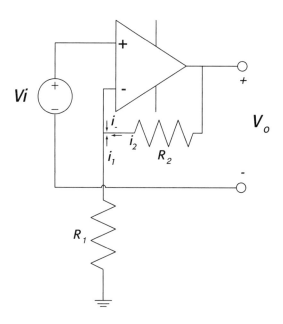

FIGURE 6.13 The noninverter configuration of an op-amp.

With the basic configurations of op-amps in place, we are ready to design some simple building block analog filters using resistors, capacitors, and op-amps.

Active *RC* Filters

The design of the analog filters requires establishing the loop equations of the four basic types of the filter Transfer Functions, namely, the low-pass, high-pass, band-pass, and band-stop filters. The requirements are specified in terms of the cutoff frequencies, establishing the poles and zeros of the filter's Transfer Function and improvement is achieved by cascading several blocks in a series. The network loop equation parameters directly correspond to the component values, and the design requirement is simply a matter of choice between the different component compositions. In 1955, R. P. Sallen and E. L. Key described these filter circuits that have become the de facto standard for the filter designs.

The circuit shown in Figure 6.14 is a generalized form of the Sallen-Key circuit, where generalized impedance terms, Z, are used for the passive filter components, and R3 and R4 are the noninverting op-amp gain.

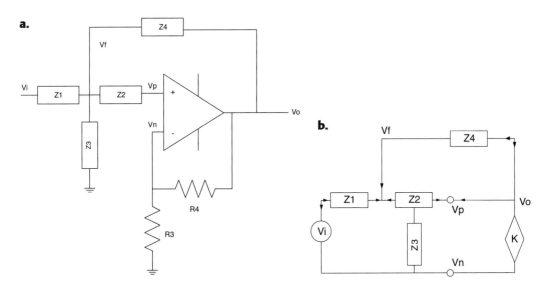

FIGURE 6.14 a) The generalized Sallen-Key circuit. b) Loops for node current analysis.

The loop analysis can be described as follows:
KCL at V_f

$$V_f(\frac{1}{Z1}+\frac{1}{Z2}+\frac{1}{Z4}) = V_i(\frac{1}{Z1})+V_p(\frac{1}{Z2})+V_o(\frac{1}{Z4})$$

KCL at V_p

$$V_p(\frac{1}{Z2}+\frac{1}{Z3}) = V_f(\frac{1}{Z2})$$

$$V_f = V_p(1+\frac{Z2}{Z3})$$

Solving for V_p

$$V_p = V_i\left(\frac{Z2Z3Z4}{Z2Z3Z4+Z1Z2Z4+Z1Z2Z3+Z2Z2Z4+Z2Z2Z1}\right)+$$
$$V_o\left(\frac{Z1Z2Z3}{Z2Z3Z4+Z1Z2Z4+Z1Z2Z3+Z2Z2Z4+Z2Z2Z1}\right)$$

The op-amp gain K

$$V_n = V_o(\frac{R3}{R3+R4})$$

Since op-amp amplifies the difference between the two input terminals,

$$V_o = K(V_p - V_n)$$

We can obtain the following generalized transfer function,

$$\frac{V_o}{V_i} = \left(\frac{K}{\frac{Z1Z2}{Z3Z4}+\frac{Z1}{Z3}+\frac{Z2}{Z3}+\frac{Z1(1-K)}{Z4}+1}\right)$$

Please see the following link for design guidelines: *http://focus.ti.com/lit/an/sloa024b/sloa024b.pdf*.

Single Pole Low-Pass Filter

The circuit of Figure 6.15 is a single pole low-pass filter, corresponding to the Transfer Function of Equation 6.10. The output V_i measured on capacitor C is given as

$$H(s) = \frac{V_O}{V_I} = \frac{Z_C}{Z_R + Z_C}$$

Substituting for the values of $Z_C = 1/j\omega C$ and $Z_R = R$, we get Equation 6.10.

$$H(j\omega) = \frac{V_O}{V_I} = \frac{1}{1 + j\omega RC} \qquad (6.10)$$

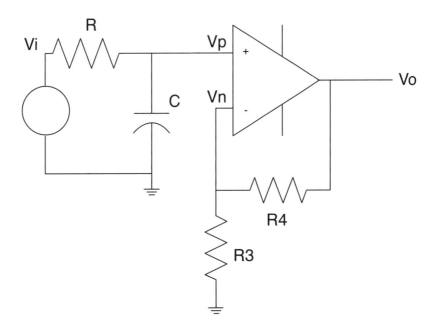

FIGURE 6.15 First order low-pass active *RC* filter.

Design Considerations

From Equation 6.10, we can draw the following filter characteristics:
The cutoff frequency is shown in Equation 6.11.

$$\omega_c = 1/RC \qquad (6.11)$$

The frequency response magnitude is shown in Equation 6.12.

$$|H(j\omega)| = \frac{1}{RC} \times \frac{1}{\sqrt{(1/RC)^2 + \omega^2}} \qquad (6.12)$$

The gain in decibels is shown in Equation 6.13.

$$20\log|H(j\omega)| = 20\log|\frac{1}{RC}| - 20\log|\sqrt{(1/RC)^2 + \omega^2}| \qquad (6.13)$$

The phase change:

$$\Phi = -\arctan(\omega RC)$$

A plot of the frequency response magnitude and phase change is presented in Figure 6.6.

EXAMPLE 6.1

Design a low-pass filter with the cutoff frequency of $\omega = 2\pi 1000\,rad/s^{-1}$.
From Equation 6.10, we have

$$1/RC = 2\pi 1000\,rad/s^{-1}$$

Since we have two element values to choose from, by selecting $R = 10\,K$, we get

$$C = 1/2\pi 1000 * 10000 = 1.57 \times 10^{-7}\,\mu F$$

EXAMPLE 6.2

Define the cutoff frequency for $R = 10\,K$ and $C = 0.001$ in Figure 6.14.
From Equation 6.11,

$$\omega_c = 1/(10 \times 10^3 \times 0.001 \times 10^{-6}) = 100\,krad/\sec$$

EXAMPLE 6.3

Define the magnitude in dB at frequency 10 KHz, for $R = 1K$ and $C = 0.01 \mu F$ in Figure 6.1.

$$\omega = 2\pi f = 2\pi \times 10^4 \, rad/s$$

From Equation 6.11,

$$1/RC = 1/10^{-8} \times 10^3 = 10^5 \, rad/sec$$

From Equation 6.12,

$$|H(j\omega)| = \frac{10^5}{\sqrt{(2\pi \times 10^4)^2 + (10^5)^2}} = 0.97$$

From Equation 6.13,

$$\text{Magnitude in dB } 20 \log(0.97) = -0.26 dB$$

Second Order Low-Pass Filter

The roll-off rate could be improved to 40 dB with the implementation of a second order circuit, as shown in Figure 6.16. The cutoff frequency ω_c is normalized to 1 so the design could be standardized, and only in the final analysis we scale up the component values to represent the true cutoff point. The generalized Transfer Function is given as shown in Equation 6.14.

$$H(s) = \frac{k\omega_C^2}{s^2 + (\omega_C/Q)s + \omega_C^2} \qquad (6.14)$$

For a second order Butterworth response

$$\omega_C/Q = \sqrt{2}$$

$N = 2$

$$s_1 = \cos(\pi 3/4) + j\sin(\pi 3/4) = -.707 + j.707$$

$$s_2 = \cos(\pi 5/4) + j\sin(\pi 5/4) = -.707 - j.707$$

$$H(s) = \frac{1}{(s_2^2 + \sqrt{2}s_2 + 1)}$$

The loop equation is given as shown in Equation 6.15.

$$H(s) = \frac{V_o}{V_I} = \frac{K/R_1C_1R_2C_2}{s^2 + s(1/R_1C_1 + 1/R_2C_1 + 1/R_2C_2 - K/R_2C_2) + 1} \qquad (6.15)$$

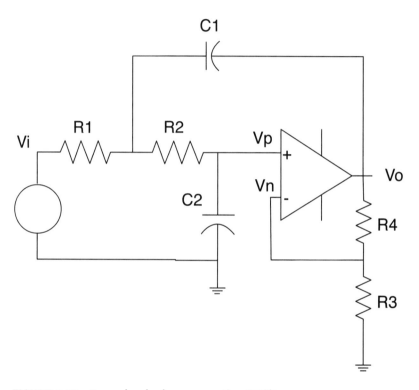

FIGURE 6.16 Second order low-pass active *RC* filter. Source: M.S. Ghansi, Electronic Devices and Circuits, (HRW, 1985).

EXAMPLE 6.4

Design a Butterworth filter with a −3 dB gain at $2\pi 10000 rad/$sec. The circuit of Figure 6.15 has input resistor $R_1 = 1K$.

Equating the Transfer Function of Equation 6.14 and the loop Equation 6.15, we get three equations (Equations 6.16 through 6.18) and five unknowns R_1, R_2, $C1$, C_2, and K,

$$H(0) = 10 = K/R_1C_1R_2C_2 \tag{6.16}$$

$$\sqrt{2}\omega_C = (1/R_1C_1 + 1/R_2C_1 + 1/R_2C_2 - K/R_2C_2) \tag{6.17}$$

$$\omega_C^2 = 1/R_1C_1R_2C_2 \tag{6.18}$$

Let's use normalized values of $R_1 = 1$, $C_1 = 1$, and $\omega_c = 1$, and solve for the other unknowns using Equations 6.16, 6.17, and 6.18.

$$H(s) = \frac{1}{s^2 + (1/Q)s + 1} = \frac{1}{(s^2 + (2/C_1)s + 1)C_1C_2}$$

$$C_1 = 2Q$$

$$C_2 = \frac{1}{2Q}$$

To realize a second order, we equate the coefficients,

$$s^2 + (1/Q)s + 1 = s^2 + \sqrt{2}s + 1$$

$$Q = 1.414$$

Using the resistor value $R_1 = 1K$ and the frequency scaling $2\pi 10000 rad/\text{sec}$, multiply each capacitor by $\frac{1}{1000} \times \frac{1}{2\pi 10000}$.

$$C1 = \frac{2 \times 1.414}{1000 \times 2\pi 10000} = .045 \mu F$$

$$C2 = \frac{1}{2 \times 1.414 \times 1000 \times 2\pi 10000} = .0056 \mu F$$

$$R_1 = 1K \qquad R_2 = 1K$$

The Band-Pass Filter

All resonant circuits act as band-pass filters. The circuit of Figure 6.16 corresponds to the Transfer Function of Equation 6.19, which allows a certain band of frequencies to pass through while suppressing the rest.

$$H(s) = \frac{H_2 s}{s^2 + (\omega_0/Q)\omega_0 + \omega_0^2} \tag{6.19}$$

There is one frequency (the resonant frequency, also called the center frequency) that has the maximum output voltage V_{max}. The frequency to the left of the center frequency with the magnitude $0.707\ V_{max}$ is the low cutoff frequency, and the frequency to the right with $0.707 V_{max}$ is the high cutoff frequency. The bandwidth is between the low and high cutoff points.

The loop equation of the circuit of Figure 6.16 is given as shown in Equation 6.20.

$$H(s) = \frac{V_o}{V_I} = \frac{-(s/C_1 R_1)}{s^2 + s(C_1 + C_2)/C_1 C_2 R_2 + 1/C_1 R_1 C_2 R_2} \tag{6.20}$$

Equating the parameters of Equations 6.19 and 6.20 and simplifying Equation 6.20 by letting $C_1 = 1$, $C_2 = 1$, and $R_2 = 1$, we can reduce the choice to just one variable R_1 to be specified,

$$H(s) = \frac{V_o}{V_I} = \frac{-(s/R_1)}{s^2 + 2s + 1/R_1}$$

The cutoff frequency is shown in Equation 6.21.

$$\omega_0^2 = 1/R_1 \tag{6.21}$$

The bandwidth quality factor is shown in Equation 6.22.

$$(\omega_0/Q) = 2$$
$$Q = 1/\sqrt{R_1} \tag{6.22}$$

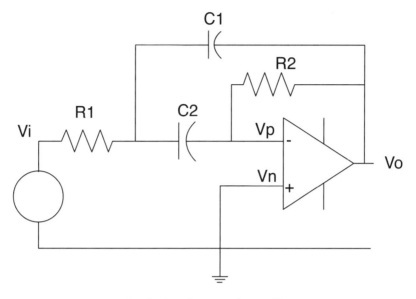

FIGURE 6.17 Second order band-pass active *RC* filter. Source: M.S. Ghansi, Electronic Devices and Circuits, (HRW, 1985).

EXAMPLE 6.5

Design a band-pass filter with $Q = 2$ and $\omega = 2\pi 1000 \, rad/s^{-1}$ with $R_1 = 10K$.

The choice could be narrowed down by making $C_1 = C_2 = C$ and $R_1 = R_2 = R$ and substituting the values into Equations 6.21 and 6.22, resulting in Equations 6.23 and 6.24.

$$\omega_0^2 = 2/(CR)^2 = [2\pi 1000]^2 \qquad (6.23)$$

$$(\omega_0/Q) = (1/CR) = 2\pi 1000 \qquad (6.24)$$

Let $R = 10K$ and from Equations 6.23 and 6.24 we get

$$C_1 = C_2 = 1.6 \times 10^{-6} \, \mu F \qquad R_1 = R_2 = 10K$$

Single Pole High-Pass Filter

Implementing the Transfer Function of Equation 6.3 gives a single pole high-pass filter. Notice that swapping the resistor and capacitor of the low-pass filter of Figure 6.14 converts it into a high-pass filter, as shown in Figure 6.17.

The loop equation is described as shown in Equation 6.25.

$$H(j\omega) = \frac{V_O}{V_I} = \frac{1}{1 - j(1/\omega RC)} \quad (6.25)$$

Design Considerations

From Equation 6.25, we can draw the following filter characteristics,
The cutoff frequency is shown in Equation 6.26.

$$\omega_c = 1/RC \quad (6.26)$$

The frequency response magnitude is shown in Equation 6.27.

$$|H(j\omega)| = \frac{1}{RC} \times \frac{1}{\sqrt{(1/RC)^2 + \omega^2}} \quad (6.27)$$

The gain in decibels is shown in Equation 6.28.

$$20 \log |H(j\omega)| = 20 \log \left|\frac{1}{RC}\right| - 20 \log \left|\sqrt{(1/RC)^2 + \omega^2}\right| \quad (6.28)$$

The phase change is shown in Equation 6.29.

$$\Phi = -\arctan(1/\omega RC) \quad (6.29)$$

A plot of the frequency response is presented in Figure 6.18.

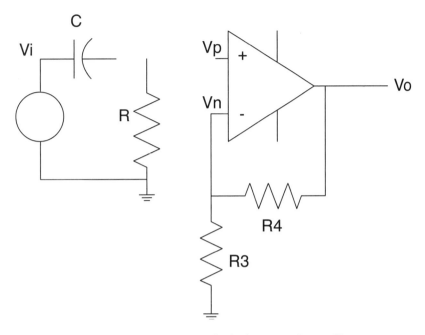

FIGURE 6.18 The circuit for a first order high-pass active RC filter. Source: M.S. Ghansi, Electronic Devices and Circuits, (HRW, 1985).

EXAMPLE 6.6

Design a high-pass filter with cutoff frequency of $\omega = 2\pi 1000 \, rad/s^{-1}$ and the transition band of 20 dB per decade.

The single pole solution of Equation 6.25 satisfies 20 dB requirements.
Substituting the cutoff frequency $\omega = 2\pi 1000 \, rad/s^{-1}$ into Equation 6.26, we get

$$(1/RC) = 2\pi 1000$$

If we pick $R = 1K$, the capacitor $C = 1.57 \times 10^{-6} \mu F$

EXAMPLE 6.7

Define the magnitude and phase of the filter output of Example 6.6 at 100 Hz.

The magnitude from Equation 6.27:

$$|H(j\omega)| = 2\pi 1000 \times \frac{1}{\sqrt{(2\pi 1000)^2 + 100^2}} = 0.999$$

The log magnitude from Equation 6.28:

$$20 \log |0.999| = -0.01$$

The phase from Equation 6.28:

$$\Phi = -\arctan(2\pi 100 \times 2\pi 1000) = 0.001 \deg$$

Second Order High-Pass Filter

Implementing the Transfer Function of Equation 6.30 provides a 40 dB roll-off rate.

$$H(s) = \frac{H_2 \omega}{\omega^2 + b_1 \omega + b_0} \quad (6.30)$$

The circuit of Figure 6.19 has the loop equation matching the Transfer Function

$$H(s) = \frac{V_o}{V_I} = \frac{Ks^2}{s^2 + s(1/R_1 C_1 - K/R_2 C_1 + 1/R_2 C_2 + K/R_2 C_2) + 1/R_1 C_1 R_2 C_2} \quad (6.31)$$

There are five unknowns—R_1, R_2, C_1, C_2, and the gain K—with three equations:

The gain

$$H = K$$

The cutoff frequency

$$\omega_c^2 = b_o = \frac{1}{R_1 R_2 C_1 C_2}$$

For a second order Butterworth response

$$b_1 = \sqrt{2}\omega_C = \frac{1-K}{R_1 C_1} + \frac{1}{R_2 C_2} + \frac{1}{R_2 C1}$$

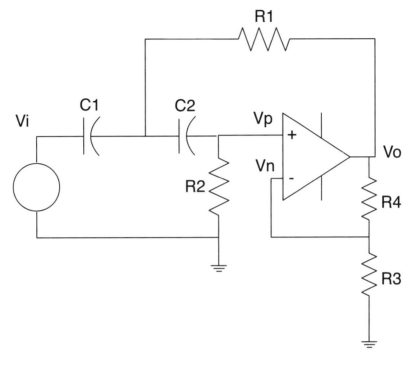

FIGURE 6.19 The circuit for a second order high-pass active *RC* filter.

EXAMPLE 6.8

Design a second order high-pass Butterworth filter with a −3 dB gain at $2\pi 10000 rad$/sec and a gain of 10 at $2\pi 10000 rad$/sec dB. The circuit of Figure 6.19 has the resistor value $R_1 = 1K$.

Equating the Transfer Function of Equation 6.30 and the loop of Equation 6.31, we get three equations and five unknowns: R_1, R_2, C_1, C_2, and K. We could normalize the values of $R_1 = 1, C_1 = 1, \omega_0 = 1$ and obtain the three unknown values in normalized form:

$$K = 10 \qquad C_2 = 9.414 \qquad R_2 = 9.414 \qquad R_1 = 1 \qquad C_1 = 1$$

The actual values are

$$R_1 = 1K \qquad R_2 = 1K \times 0.11 = 111\Omega \qquad C_1 = .016\mu F \qquad C_2 = .15\mu F$$

Band-Stop Filter

A band stop response is achieved by canceling the band-pass response with a pure resonant response from the numerator polynomial.

The circuit shown in Figure 6.20 has a gain, as shown in Equation 6.32.

$$H(s) = \frac{H_4(s^2 + \omega^2)}{s^2 + (\omega_p/Q)\omega_0 + \omega_p^2} \tag{6.32}$$

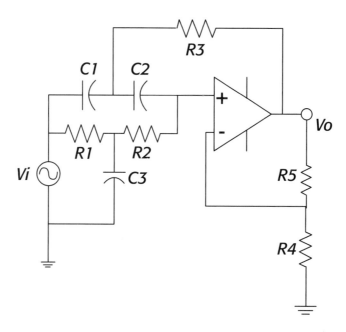

FIGURE 6.20 Circuit for a second order band-stop active *RC* filter. Source: M.S. Ghansi, Electronic Devices and Circuits, (HRW, 1985).

The loop equations are given as

$$H(s) = \frac{V_o}{V_I} = \frac{K(C_1 C_2 s^2 + 1/R_1 R_2)}{C_1 C_2 s^2 + s[(C_1 + C_2)/R_2 + C_2(1/R_1 + 1/R_2)(1-K)] + 1/R_1 R_2}$$

There are five design parameters, R_1, R_2, C_1, C_2, and the gain K and three equations, making $C_1 = C_2 = C$ and $R_1 = R_2 = R$, we can reduce the number of variables to three and solve for them using,

$$H(s) = \frac{V_o}{V_I} = \frac{K(C^2 s^2 + 1/R^2)}{C^2 s^2 + 2Cs/R + 1/R^2}$$

SUMMARY

In this chapter, we studied the filter design methods of the four basic types of analog filters: the low-pass, high-pass, band-pass, and band-stop filters. The filters were designed as solutions of differential equations that governed the input and output relationship forming the filter transfer function and implemented as electrical circuits of resistors, capacitors, and op-amps whose component values matched the coefficient values of the transfer function. The poles and zeros of the transfer functions describe the characteristics of the filters such as the cutoff frequency and the roll-off rate in the transition band, and the filter response was measured by how well the filter suppresses frequencies beyond the cutoff region. It is possible to improve upon the filter response by adding extra poles at the cutoff point that produce steeper roll-off in the filter transition band.

7 Digital Filters

In This Chapter

- The Digital Filter Design Process
- Digital Filter Characterization
- Realizing the Filter Response
- Improving the Filter Response
- Butterworth Filters
- Chebyshev Filters
- Elliptic Filters
- The Transition from Analog to Digital Filters

Energy storage elements produce frequency response that acts as filters, and we have seen in the analog filters how different combination of resistors, capacitors, and op-amps exhibit the filter action. The same effect can be achieved using computers in the form of digital filters, but there is much more to it than just following the analog response. Some of the techniques in the digital filters have no parallel in the analog design, such as the sampled data processing of averaging and windowing that is only possible in the digital domain. Although the digital filters are diversified in nature, from the control system point of view, replacing the analog filters with the counter part digital filters is an important consideration in any system design.

A filter is generated when a delayed input and or output is fed back to the system after being multiplied by a complex quantity (see Figure 7.1). The complex quantity multiplier forms the Transfer Function of the system, and how to formulize it is the topic of discussion in this chapter.

The analog filters of the previous chapter produced the frequency response through implementing a Transfer Function, whose coefficient values matched the component values. But the digital filters will be implemented as a solution of the difference equation (obtained from the corresponding Transfer Function), the output of which will be a discrete time data. The coefficients would still be obtained from the corresponding frequency response Transfer Function.

Appendix C, "Digital Filter Applications," is provided with an application program "filterTest.pl" that may be used as an aid to design and analyze the response of simple pole and zero filters. The program is based on free plotting software (xmgrace) and the perl Tk module. In the first part of this chapter, we will devote our attention to the digital filters that are derived from the analog designs and require feedback from the previous output. Later we will discuss the sample data processing that does not require any feedback from the previous output.

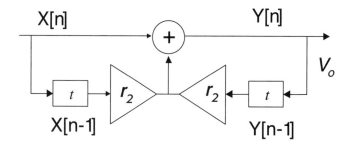

FIGURE 7.1 The Block diagram of digital filters, with feedback and feed forward mechanisms, where r_1 and r_2 are the complex number multipliers and $x[n-1]$ and $y[n-1]$ are the previous inputs and outputs.

DIGITAL FILTER DESIGN PROCESS

Filters are identified by the frequency response they produce. The low-pass, high-pass, band-pass, and band-stop responses are the basic types that characterize a filter. With each frequency response, there is a corresponding Transfer Function describing the input and output relationship. The analog domain Transfer Function is the Fourier Transform of the continuous time impulse response, whereas the digital domain Transfer Function is the z-Transform of the impulse response. The

Fourier Transform is a function of the continuous time frequency variable ωt and the z-Transform is a function of the discrete time frequency variable $\omega_{k\Delta t}$. The analog filters generate the output as if there is a continuous time convolution; on the other hand, the digital filter's output is from the discrete time convolution, as you will see later in this section.

In the next derivative, you'll see that a filter's frequency response to a given input (essentially the Fourier Transform of the impulse response times the input function) is the same as a convolution integral in the time domain. Once we establish that the output of a system can be obtained by simply convolving the input with its impulse response, the digital signal processing may be implemented as an iterative algorithm of additions of input multiplied with the impulse response.

The Convolution Sum

For a system with an impulse response $h(t)$ and the input $x(t)$, the convolution is given as shown in Equation 7.1.

$$y(t) = \int_{-\infty}^{\infty} h(\tau) x(t - \tau) d\tau \qquad (7.1)$$

For a sinusoidal input e^{st}, the output is given as

$$y(t) = \int_{-\infty}^{\infty} h(\tau) e^{s(t-\tau)} d\tau$$

$$y(t) = e^{st} \int_{-\infty}^{\infty} h(\tau) e^{-s\tau} d\tau$$

The integral on the right-hand side of Equation 7.1 is the Fourier Transform of the impulse response and may be written as the Transformed output shown in Equation 7.2.

$$y(t) = H(s) e^{st} \qquad (7.2)$$

Equation 7.2 being derived from Equation 7.1 indicates that the convolution of the impulse response with the input signals is equivalent to producing the frequency response of the Transformed output.

The discrete time version of the convolution process is given in Equation 7.3.

$$y[n] = \sum_{k=-\infty}^{\infty} h[n-k] \times x[k] \qquad (7.3)$$

The same result could be achieved when we change the order of convolution; Equation 7.3 could also be written like Equation 7.4.

$$y_n = \sum_{k=-\infty}^{\infty} x[n-k] \times h[k] \qquad (7.4)$$

There is a subtle difference between Equations 7.3 and 7.4. The convolution of the impulse response with the input signals (Equation 7.3) requires a complete set of input data before the convolution can begin, whereas Equation 7.4 produces the output as soon as the first input signal is available and this is the preferred method for digital signal processing applications. Limiting the impulse response to a finite length may eliminate the infinite sum. Thus, for an impulse response with N number of terms, the convolution sum is given as shown in Equation 7.5.

$$y[n] = \sum_{k=0}^{N} x[n-k] \times h[k] \qquad (7.5)$$

The convolution process of Equation 7.5 may realize a digital filter, but the Transfer Function of its impulse response describes its characteristics such as the frequency and the phase response. Before we proceed further with the discussion of Transfer Function, the concept of sampling frequency should be clarified.

The Frequency Variable $\omega \Delta t$

There is a certain amount of time delta between successive scanning of input signals by digital computers. The input data taken goes through the processing of multiplying with the complex number Transfer Function before the next data is scanned, and this processing delay is Δt. The value of the continuous time input function is valid only at the time $k\Delta t$, where k is the sample number (see Figure 7.2). Thus, we have a frequency variable that varies with the sample number instead of sample time.

The term $\omega \Delta t$ will be used quite often in the digital filter design, so let's reserve the capital Ω specifically for indicating the sampling frequency variable,

$$\omega \Delta t = \Omega$$
$$z^1 = e^{j\omega \Delta t} = e^{j\Omega} = \cos\Omega + j\sin\Omega$$
$$z^2 = e^{j\omega 2t} = e^{j2\Omega} = \cos 2\Omega + j\sin 2\Omega$$
$$z^{-1} = e^{-j\omega \Delta t} = e^{-j\Omega} = \cos\Omega - j\sin\Omega$$

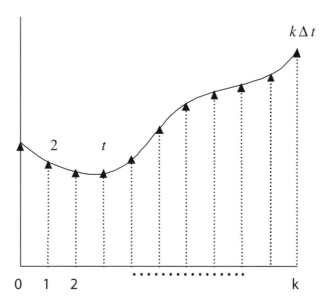

FIGURE 7.2 The continuous time and discrete time sampling correspond at $t = k\Delta t$.

The sampling frequency variable Ω should be seen in the context of the Nyquist frequency, which is half of the sampling frequency. For example, for a sampling rate of 1,000 samples per second, the Nyquist frequency is 500 cycles per second, or $2\pi 500$ radians per second. The cutoff frequency parameters for all basic filter design functions are normalized by the Nyquist frequency, so if you are designing a filter for a 1,000 Hz frequency and want to suppress frequencies beyond 100 Hz ($2\pi 100$ radians/sec), the corresponding cutoff point in terms of pole frequency is 0.2Ω *radians/sample*. Another way of looking at the sampling frequency is the time delay Δt. Suppose you have a data acquisition system that scans input values every 1 msec. The Nyquist frequency requires minimum of 2 samples, then the highest frequency the system can process is 500 Hz, since $f = 1/2\Delta t = 1/.002 = 500 Hz$. Computers only see frequencies up to the Nyquist frequency, which is different from with analog domain filters where the range is unlimited. In the digital domain, the Nyquist frequency is the highest frequency that we can pass through a digital filter. (We start seeing the alias effect after that.) If you suspect there are higher frequency components in your system beyond the Nyquist frequency, you must add a hardware filter before passing data through a digital filter.

Digital Filter Transfer Functions

In a linear and time invariant system, an input signal of the form (Ae^{-st}) when multiplied with a complex number $H(s)$ produces a new value of the same frequency but a different magnitude and phase, and that is what a filter outcome is; the magnitude and phase of the input signal may be altered but the frequency remains the same. A low-pass filter shortens the magnitude of higher frequencies, whereas a high-pass filter shortens the magnitude of the lower frequencies; we discussed this in Chapter 5, "Laplace Transforms and z-Transforms," from the analog domain point of view, but digital filters are different, so their Transfer Function will be discussed from a different context.

To visualize the effect of a digital filter Transfer Function, imagine a vector whose magnitude varies with the frequency and it is multiplied by the input signal. The normalized sampling frequency function $z = e^{j\Omega}$ is a vector, rotating counterclockwise from $\Omega = 0$ to the maximum of π radians per sample, reaching the Nyquist frequency. Figure 7.3 depicts the magnitude of such a vector at various points on the semicircle. Each division on the semicircle is a normalized sampling

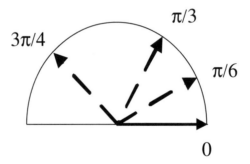

FIGURE 7.3 The Transfer Function composed of a single vector z, rotating counterclockwise. Notice the magnitude remains the same throughout the sampling frequency range of $\Omega = 0$ to $\Omega = \pi$.

frequency. If z is the only vector in our Transfer Function, there is no gain and no loss, as there is no change in the output.

But if we add a constant vector $re^{j\theta}$ to the vector z, it becomes a distance vector $(z - re^{j\theta})$, which is a function of frequency variable Ω. If we plot the distance of this vector to the unit circle, we see its length varies with the frequency as shown by the thin arrows in Figures 7.4 and 7.5. The length indicates the magnitude of vector $z - re^{j\theta}$ as it travels from 0 frequency to the frequency of π radians per sample. The

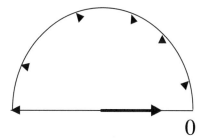

FIGURE 7.4 The magnitude of the vector $(z - r)$, $(re^{j0} = r)$ as it goes through different sampling frequencies. Notice the increase in magnitude as the vector $(z - r)$ rotates from $\Omega = 0$ to $\Omega = \pi$.

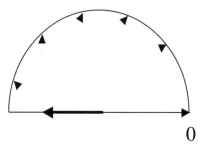

FIGURE 7.5 The magnitude of the vector $(z - r)$, $(re^{j180} = -r)$ as it goes through different sampling frequencies. Notice the decrease in magnitude as the vector $(z - r)$ rotates from $\Omega = 0$ to $\Omega = \pi$.

magnitude increases if $\theta = 0°$ and decreases if $\theta = 180°$, as it travels towards the other end, near the π radians per sample.

Suppose there is an input signal containing all frequencies of equal magnitude and is multiplied by the magnitude of the vector $(z - r)$ of Figure 7.4. The result would have the higher frequency amplified compare to the lower frequency (a high-pass filter). The effect would be opposite if we were to divide the input; the system would act like a low-pass filter. This is just a crude example; in reality, you may need a complex Transfer Function with several of such vectors to get the desired filter affect, as you will see later in the chapter.

The Transfer Function Frequency and Phase Response

The magnitude of a simple Transfer Function (such as the vector $(z - r)$) may be obtained by multiplying with its complex conjugate, as shown in Equation 7.6.

$$|z| = \cos^2 \Omega + \sin^2 \Omega = 1$$
$$|H(\Omega)| = |z - r| = \sqrt{(\cos \Omega - r) + (j \sin \Omega) \times (\cos \Omega - r) - (j \sin \Omega)} \quad (7.6)$$

Simplifying Equation 7.6 using the trigonometric identity $\cos^2 \theta + \sin^2 \theta = 1$, we get Equation 7.7.

$$|H(\Omega)| = |z - r| = \sqrt{1 - 2r \cos \Omega + r^2} \quad (7.7)$$

The frequency response of the Transfer Function may be obtained by evaluating Equation 7.7 as a function of frequency variable Ω.

The phase response as a function of frequency may be computed as,

$$\Phi = \arctan \frac{\sin \Omega}{\cos \Omega - r}$$

The two aspects of the filter design we are interested in are its frequency and phase response. The *frequency response* describes how the magnitude changes as a function of frequency, and the *phase response* describes the delay in the output for a specific frequency response.

Transfer Function Models

A Transfer Function may be represented in several different ways; but we will discuss only two: the rational polynomial form and the pole-zero-gain form.

Rational Polynomial Transfer Functions

Suppose we have a feedback system where the output is fed back to the system after a delayed time and multiplied by a complex constant $re^{j\theta}$ as follows,

$$y_n = x_n + re^{j\theta} y_{n-1}$$

For $\theta = 0°$, the constant multiplier $re^{j\theta}$ is a real positive value ($re^{j0} = r$), and for $\theta = 180°$ it is a real negative value ($re^{j180} = -r$). The delayed output is expressed as,

$$y_{n-1} = re^{j(k-1)\omega \Delta t} = re^{jk\omega \Delta t} e^{-j\omega \Delta t}$$

In the familiar Transformed notation,

$$x_n \to X \qquad y_n \to Y \qquad ry_{n-1} \to rYz^{-1}$$

The feedback system may be expressed in the form of a *rational polynomial* of numerator and denominator as shown in Equation 7.8.

$$Y(1 - rz^{-1}) = X$$

$$H(\omega \Delta t) = \frac{Y}{X} = \frac{1}{(1 - rz^{-1})} \qquad (7.8)$$

Equation 7.8 simply states that the denominator polynomial of the system's Transfer Function multiplies the system's output, and the delayed output is fed back to the system.

A different situation may arise in a feed forward system such as,

$$y_n = x_n - rx_{n-1}$$

The delayed input may be expressed as,

$$x_{n-1} = re^{j(k-1)\omega \Delta t} = re^{jk\omega \Delta t} e^{-j\omega \Delta t}$$

In the familiar Transformed notation,

$$x_n \to X \qquad y_n \to Y \qquad x_{n-1} \to rXz^{-1}$$

The input and output relationship may be expressed as shown in Equation 7.9.

$$Y = X(1 - rz^{-1})$$

$$H(\omega \Delta t) = \frac{Y}{X} = (1 - rz^{-1}) \qquad (7.9)$$

Equation 7.9 states that the numerator polynomial of the system's Transfer Function multiplies the system's input, and a delayed input is fed back to the system.

The generalized form of a z-domain rational polynomial with numerators and denominators is specified as shown in Equation 7.10.

$$H(z) = \frac{Y(z)}{X(z)} = \frac{b(1) + b(2)z^{-1} + \cdots + b(nb+1)z^{-nb}}{a(1) + a(2)z^{-1} + \cdots + a(nb+1)z^{-na}} \qquad (7.10)$$

Equation 7.10 is a difference equation in z-domain, representing the Transfer Function of the impulse response of the digital filter.

Transfer Functions with Poles-Zero-Gain

Let's analyze Equations 7.8 and 7.9 by redefining the polynomials, so we can see where the roots are; see Equations 7.11 and 7.12.

$$H(z) = (1 - rz^{-1}) = \frac{z - r}{z} \qquad (7.11)$$

$$H(z) = \frac{1}{(1 - rz^{-1})} = \frac{z}{z - r} \qquad (7.12)$$

If the vector $z = re^{j\theta}$ in Figures 7.8 and 7.9 is the pole vector, its contribution to the magnitude is inversely proportional to its distance from the circle perimeter, and if it is the zero vector, its contribution is directly proportional to its distance from the same perimeter. The response has a peak at the shortest distance from the edge of the circle. In either case, it acts as a frequency discriminator. When $\theta = 0°$, the vector $z - re^{j\theta}$ lies on the right side of the semicircle and has a higher magnitude near the Nyquist frequency of π radians per sample. Being a pole vector, the net effect is the suppression of the higher frequencies from the input signals compared to lower frequencies, creating a low-pass filter. The Transfer Function represented as poles and zeros describes the filter's magnitude at certain frequencies, which may be used to characterize its frequency response.

Let's analyze the frequency response of a pole vector when $r = 0.48$.

The Transfer Function in the pole zero form is shown in Equation 7.13.

$$H(z) = \frac{Y(z)}{X(z)} = \frac{z}{z - 0.48} \qquad (7.13)$$

The magnitude as a function of frequency variable Ω is shown in Equation 7.14.

$$H(z) = \frac{1}{(-0.48)^2 - 0.48 \times 2\cos(\Omega) + 1} \qquad (7.14)$$

Figure 7.6 shows the magnitude and phase of the Equation 7.14.

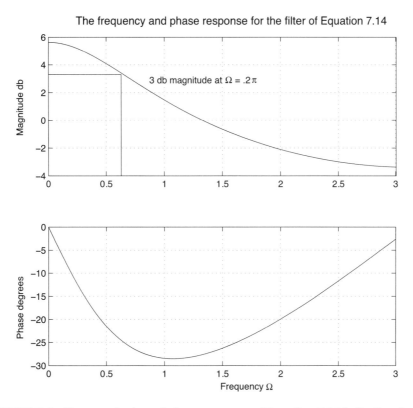

FIGURE 7.6 The magnitude and phase response of Equation 7.14 indicating the drop in magnitude to 0.707 at the sampling frequency of 0.2.

Gain

It is desirable to restrict the output to a maximum gain of unity at the peak value. A gain factor may be added to bring down the high magnitude of a pole vector. Considering the previous example, a unity gain at 0 Hz may be computed as follows.

The maximum gain at 0 Hz:

$$G = \left| \frac{e^{j\Omega}}{e^{j\Omega} - 0.48} \right| = 1.92$$

To bring the gain down to 1 at 0 Hz, the Transfer Function must be multiplied by a gain value of 1/1.92, or 0.52.

The Transfer Function described in Equations 7.11 and 7.12 is of the pole-zero-gain form. Equation 7.11 has a zero in the numerator at r and a pole in the denominator at 0, whereas Equation 7.12 has a pole in the numerator at *r* and a zero in the denominator at 0.

The rational polynomial form of the Transfer Function describes the difference equation of the system, and pole-zero-gain form describes the magnitude response of the system. Filters are realized with the polynomial form and characterized by the pole and zero form.

It should be noted that any pole frequency vector lying outside the unit area is a sign of an unstable filter. If $r > 1$, every subsequent output will be greater than the previous one, resulting in an exponential growth without bound, thus the value of $r > 1$ should be avoided.

The simple poles and zeros of Equations 7.11 and 7.12 are only the building blocks that form a filter; a complex system may have several poles and zeros, and the overall system response is a linear product of the individual response.

The factored or zero-pole-gain form of a Transfer Function is shown in Equation 7.15.

$$H(z) = \frac{q(z)}{p(z)} = k\frac{(z-q(1))(z-q(2))\cdots(z-(q(n))}{(z-p(1))(z-p(2))\cdots(z-(p(n))} \qquad (7.15)$$

To see how Equations 7.11 and 7.12 act as filters of frequencies, see them in the context of magnitude as a function of frequency variable Ω.

Filters may be analyzed with the help of the pole and zero form of the Transfer Function but are realized with the rational polynomial form of the Transfer Function.

DIGITAL FILTER CHARACTERIZATION

The four basic types of frequency response with which the filters are characterized are the low-pass, high-pass, band-pass, and band-stop filters. The requirements are generally specified in terms of the cutoff frequency, the roll-off rate, the magnitude at a specific frequency, ripples at pass-band range, etc. We will leave the more elaborate requirements for later discussion, but next we will discuss the simple design method using poles and zeros.

Low-Pass Filters

If the pole $(z - re^{j\theta})$ lies on the right side of the semicircle ($\theta = 0$), its distance would increase as we go further towards the higher end of π radians per sample (see the

Figure 7.4), since, having an inverse effect, the Transfer Function magnitude would be lower at the higher frequencies. The net effect is suppression of the higher frequencies from the input signals, whereas the lower frequencies are amplified, essentially creating a low-pass filter.

EXAMPLE 7.1

A data acquisition system has a sampling rate of 1,000 samples per second. Formulate a single pole Transfer Function with the cutoff frequencies of 100 Hz. The Nyquist frequency:

$$1000/2 = 500 \ Hz$$

The cutoff point as a fraction of the Nyquist frequency is

$$100/500 = 0.2\Omega$$

At the cutoff frequency the magnitude drops down to 0.707. Solving for r when the magnitude is 0.707,

$$|H(\Omega)| = \left|\frac{z}{z-r}\right| = \frac{1}{\sqrt{1-2r\cos\Omega+r^2}} = .707$$

$$|H(\Omega)| = \left|\frac{z}{z-r}\right| = r^2 - 2r\cos(.2 \times \pi) - 1 = 0$$

$$r = 0.48$$

The Transfer Function in the pole zero form is shown in Equation 7.16.

$$H(z) = \frac{Y(z)}{X(z)} = \frac{z}{z-0.48} \qquad (7.16)$$

In the rational polynomial form, the Transfer Function is,

$$H(z) = \frac{Y(z)}{X(z)} = \frac{1}{1-0.48z^{-1}}$$

The magnitude as a function of frequency variable Ω is shown in Equation 7.17.

$$H(z) = \frac{1}{(-0.48)^2 - 0.48 \times 2\cos(\Omega) + 1} \tag{7.17}$$

See Figure 7.6 for the magnitude and phase response of Equation 7.17.

Matlab and Scilab to the Rescue

Filter modeling requires extensive computations and analysis such as computing the magnitudes and transforming poles and zeros to rational polynomials or finding roots of a polynomial, etc. Matlab and Scilab have built-in functions that can help you solve these problems quickly; tutorials are provided in Appendixes A, "Matlab Tutorial," and B, "Scilab Tutorial," for exercise in filter modeling, and it is recommended that you go through them and gain confidence.

We could solve the problem of Example 7.1 with the help of Matlab by first finding the roots of Equation 7.7 and using them to compute the frequency response:

```
>> r=roots([1 -2*(cos(.2*pi)) -1])
r=

-0.4773
2.0953
```

Entering the numerator polynomial:

```
>> b=[1 0];
```

Entering the denominator polynomial:

```
>> a=[1 -0.4773];
```

Get the frequency response as shown in Figure 7.6:

```
>> freqz(b,a);
```

High-Pass Filter

If the pole lies on the left side of the semicircle ($\theta = 180°$), its distance would increase as we go further towards the lower end of 0 radians per sample (see Figure 7.8), since, having an inverse effect, the Transfer Function magnitude would be lower at the lower frequencies. The net effect is suppression of the lower frequencies from the input signal (a high-pass filter).

EXAMPLE 7.2

Going back to the previous example, we can change the low-pass filter to a high-pass filter by simply changing the phase angle of the vector from 0° to 180°; this would give us the magnitude of r as

$$r = 0.48e^{-j180} = -0.48$$

The rational polynomial Transfer Function is shown in Equation 7.18.

$$H(z) = \frac{Y(z)}{X(z)} = \frac{1}{1 + 0.48z^{-1}} \qquad (7.18)$$

The maximum gain at Nyquist frequency (π radians/sample):

$$G = \frac{z}{z + 0.48} = \frac{e^{j180}}{e^{j180} + 0.48} = 1.92$$

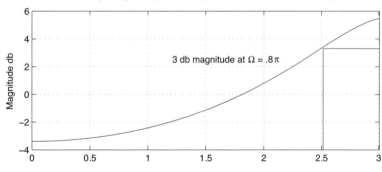

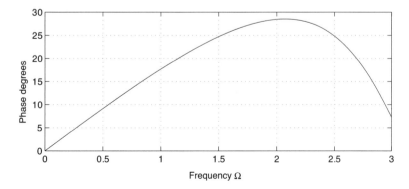

FIGURE 7.7 The magnitude and phase response of the high-pass filter of Example 7.2.

The Band-Pass and the Band-Stop Filters

The response of the real vector poles and zeros was easy to visualize as the maximum magnitude occurred only at the phase angle θ of either 0° or 180°. But the effect is different if θ is anything other than 0° or 180°. The multiplier $re^{j\theta}$ forms a complex vector if $0 < \theta < 180°$. With the complex vector $re^{j\theta}$, there is always a corresponding complex conjugate $re^{-j\theta}$, as shown in Figures 7.8a and 7.8b. The combined magnitude of the two vectors varies, as the vectors move along from 0 to π, but precisely at the phase angle (θ), the magnitude has its shortest value, resulting in the highest multiplier if it is a pole vector and lowest multiplier if it is a zero vector. This is the multiplier with which the input frequencies are multiplied. Thus, the combined magnitude of the conjugate poles and zeros creates a peak where the phase angle θ is (see Figure 7.8a for the pole effect and Figure 7.8b for the zeros effect).

In the process, the frequencies in the neighborhood of θ are either enhanced or suppressed (based on whether the vector is a pole or a zero) and such neighborhood is the *bandwidth* of the filter and is defined as the two adjacent points where the magnitude falls to 0.707 of the peak.

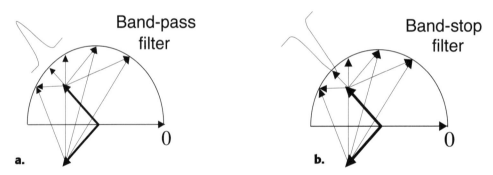

FIGURE 7.8 a) The bandwidth of a band-pass filter showing magnitude enhanced due to a complex pole vector. b) Magnitude reduced due to a complex zero vector.

The Bandwidth

To compute the bandwidth created by the vector $(1 - r)$, imagine part of the unit circle (in the neighborhood of the vector) is a straight line, forming a triangle with the three sides d, $\sqrt{2}(1-r)$, and $(1-r)$, as shown in Figure 7.9. The distance d corresponds to one half the bandwidth and is equal to the length $(1 - r)$, thus the bandwidth is

$$2d = 2(1 - r)$$

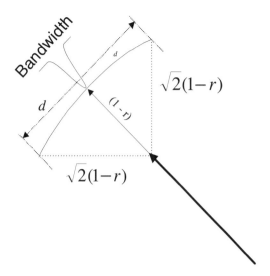

FIGURE 7.9 The bandwidth as an approximation of the triangle with the sides, d, $\sqrt{2}(1-r)$, and $(1-r)$.

Band-Pass Filters

The band-pass filters work by enhancing the magnitude of a certain range of frequencies from the rest of the inputs, as created by combined magnitude of a pair of complex conjugate pole vectors. A band-pass filter is specified by a peak frequency and the region surrounding it, where the magnitude falls to 0.707 of its peak.

Figure 7.9 describes the magnitude of the complex pole that has a peak at angle θ that tapers off on both sides. Thus, a specific band-pass frequency response may be created by strategically placing the phase angle θ in the pole vector of a filter Transfer Function. The following is the magnitude of a complex pole vector Transfer Function.

Equation 7.19 is the Transfer Function formed by complex conjugate pole vectors.

$$H(z) = \frac{1}{(1-re^{j\theta}z^{-1})(1-re^{-j\theta}z^{-1})} = \frac{z^2}{(z-re^{j\theta})(z-re^{-j\theta})} \quad (7.19)$$

$$H(z) = \frac{1}{1-2r\cos\theta z^{-1}+r^2 z^{-2}} = \frac{z^2}{z^2-2rz\cos\theta+r^2}$$

The Transfer Function magnitude:

$$|H(\Omega)| = 1/\{e^{j\omega} - re^{j\theta}\}\{e^{j\omega} - re^{-j\theta}\} = 1/e^{(2j\Omega)} - 2r\cos\theta e^{j\Omega} + r^2$$

Separating the real and the imaginary:

$$|H(\Omega)| = \frac{1}{\sqrt{(1 - 2r\cos\theta\cos\Omega + r^2\cos 2\Omega)^2 + (2r\cos\theta\sin\Omega - r^2\sin 2\Omega)^2}}$$

The phase response as a function of frequency:

$$\Phi = -\arctan\frac{(2r\cos\theta\sin\Omega + r^2\sin 2\Omega)}{(1 - 2r\cos\theta\cos\Omega + r^2\cos 2\Omega)}$$

EXAMPLE 7.3

A data acquisition system has a sampling rate of 2,048 samples per second; define the Transfer Function magnitude at the peak frequency of 300 Hz and a bandwidth of 5 Hz.

The peak frequency 300 Hz corresponds to sampling frequency:

$$\theta = \pi 300 \times 2 / 2048 = .293\pi = 52.74°$$

The bandwidth of 5 Hz corresponds to sampling frequency:

$$1 - r = d = 5/1024 = 0.00488\Omega$$

$$r = 0.995$$

The Transfer Function magnitude:

$$|H(\Omega)| = \frac{1}{\sqrt{(1 - 2r\cos\theta\cos\Omega + r^2\cos 2\Omega)^2 + (2r\cos\theta\sin\Omega - r^2\sin 2\Omega)^2}}$$

Since the peak occurs at $\theta = \Omega$, the maximum gain at the peak is

$$|H(\Omega)| = \frac{1}{\sqrt{0.0047^2 + 0.0062^2}} = 128$$

See Figure 7.10 for the response.

Gain

A unity gain at 300 Hz may be achieved by multiplying the Transfer Function with a constant value of $1/128 = 0.008$.

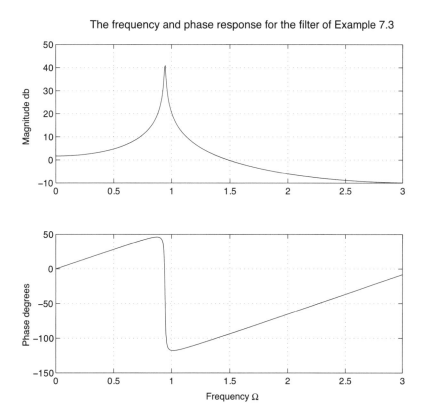

FIGURE 7.10 The magnitude and phase at different frequencies of the Transfer Function of Example 7.3.

Band-Stop Filters

As opposed to band-pass filters, the band-stop filters work by suppressing the magnitude of a certain range of frequencies from the rest of the input. Consider the same complex vector $re^{j\theta}$ with its complex conjugate $re^{-j\theta}$, but this time instead of pole, it is a zero vector of the Transfer Function.

Having a direct relationship on the input frequencies, the magnitude forms a valley precisely at the phase angle θ. Similar to the band-pass filters, the frequencies in the neighborhood of θ also get affected, but this time they are suppressed. The neighborhood is considered the two adjacent points where the magnitude is 1.414 of the lowest value.

Figure 7.12 is the response of the band-stop filter obtained from replacing the pole vector of Example 7.3 to a zero vector.

The complex vector zero magnitude is shown in Equation 7.20.

$$H(z) = (1 - re^{j\theta}z^{-1})(1 - re^{-j\theta}z^{-1}) = \frac{(z - re^{j\theta})(z - re^{-j\theta})}{z} \quad (7.20)$$

Separating the real and the imaginary, we get Equation 7.21.

$$|H(\Omega)| = \sqrt{(1 - 2r\cos\theta\cos\Omega + r^2\cos 2\Omega)^2 + (2r\cos\theta\sin\Omega - r^2\sin 2\Omega)^2} \quad (7.21)$$

The phase response as a function of frequency:

$$\Phi = \arctan\frac{(2r\cos\theta\sin\Omega + r^2\sin 2\Omega)}{(1 - 2r\cos\theta\cos\Omega + r^2\cos 2\Omega)}$$

EXAMPLE 7.4

Convert the band-pass filter of the previous example to a band-stop filter that would suppress all frequencies in the neighborhood of 300 Hz.

The frequency to suppress (300 Hz) corresponds to the sampling frequency:

$$\theta = \pi 300 \times 2 / 2048 = .293\pi = 52.74°$$

The bandwidth of 5 Hz corresponds to sampling frequency:

$$1 - r = d = 5 / 1024 = 0.00488\Omega$$

$$r = 0.995$$

Substituting the value of r and θ into the Transfer Function Equation 7.18, we obtain the desired frequency and phase response as shown in Figure 7.11.

Notch Filter

A band-pass filter made up of a single zero vector does not produce a desirable response; instead, a better result may be obtained by a complex zero of magnitude 1 to the band-pass filter to create the corresponding band-stop or notch filter, as shown in the following example.

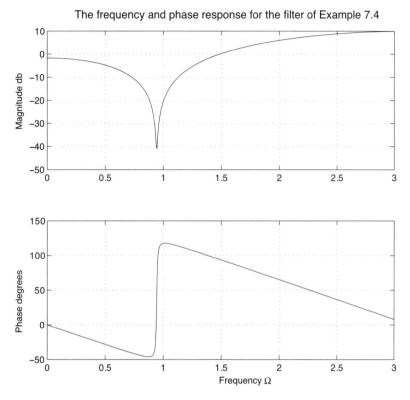

FIGURE 7.11 The magnitude and phase at different frequencies of the Transfer Function of Example 7.4.

EXAMPLE 7.5

Continuing the band-stop filter example, the peak frequency 300 Hz corresponds to sampling frequency:

$$\theta = \pi 300 \times 2 / 2048 = .293\pi$$

The bandwidth of 5 Hz corresponds to sampling frequency:

$$1 - r = d = 5 \times 2 / 2048 / 2 = 0.00288\Omega$$
$$r = 0.99756$$

The required Transfer Function with a complex pole at $r = 0.99756$ and phase $\theta = .293\pi$ and a complex zero at $r = 1$ and phase $\theta = .293\pi$ is given as

$$H(\Omega) = \frac{(z - e^{j.293\pi})(z - e^{-j.293\pi})}{(z - 0.99756 e^{j.293\pi})(z - 0.99756 e^{-j.293\pi})}$$

The Transfer Function magnitude is

$$|H(\Omega)| = \frac{\sqrt{(1 - 1.21\cos\Omega + \cos 2\Omega)^2 + (1.21\sin\Omega - \sin 2\Omega)^2}}{\sqrt{(1 - 1.20\cos\Omega + 0.995\cos 2\Omega)^2 + (1.20\sin\Omega - 0.995\sin 2\Omega)^2}}$$

See Figure 7.12 for the response.

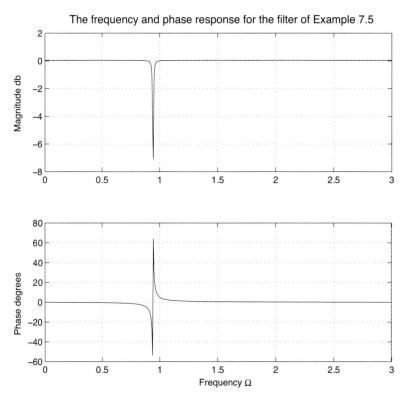

FIGURE 7.12 The magnitude and phase at different frequencies of the Transfer Function of Example 7.5.

The distance and the angle of the pole and zero vectors (with reference to the perimeter of the unit radius circle) provide some clues of what the individual vector response is going to be. Being a linear system, the product of the individual vector response is the overall filter response. A system may have several poles and zeros, but the best design is the one that achieves the objective with as few as possible.

REALIZING THE FILTER RESPONSE

Once a suitable frequency and phase response is obtained, the next step is to realize the filter in real time. A filter may be realized by multiplying the Transfer Function with the input Transform, but that requires obtaining the inverse Transform of the Transfer Function (a very complex, time-consuming process). The alternate is to convolve each input with the coefficients of the difference equation that formed the filter Transfer Function (we have shown that transform multiplication is the same as convolution).

The design of the digital filters is essentially, selecting coefficients of the governing difference equation that form the polynomial form of the filter Transfer Function. Once the coefficients are identified, it is only a matter of reading the current input, multiplying with the coefficient, and adding to the previously computed input and output, to generate the new output (essentially, solving a difference equation through convolution).

The Coefficients of the Difference Equation

Chapter 5 described the input and output relationship as a rational polynomial of frequency variables s and z. The s is the continuous time variable $e^{-j\omega t}$ and the z is the discrete time variable $e^{-j\omega k \Delta t}$. The factored form of the numerator polynomial created zeros and the denominator polynomial created poles. The factored form should be converted into a rational polynomial to realize the filter coefficients. Matlab has built-in functions to support the conversion from one form to another (see Appendix A for instructions). The rational polynomial Transfer Function can be written as shown in Equation 7.22.

$$z^0 Y(z) + a_1 z^{-1} Y(z) + ...a_m z^{-m} Y(z) = z^0 b_0 X(z) + b_1 z^{-1} X(z) + .. + b_n z^{-N} X(z) \quad (7.22)$$

Using the time shift property of z-Transform, Equation 7.22 could be written as the governing difference equation of the impulse response of the filter, as shown in Equation 7.23.

$$y_0(k) + a_1 y(k-1) + ...a_m y(k-M) = b_0 x(k) + b_1 x(k-1) + .. + b_n x(k-N) \quad (7.23)$$

To the left of Equation 7.23, we have the current output $y_0(k)$ plus all the previous outputs up to the order M of the difference equation. The right side of the equation has the current input $x_0(k)$ times its coefficient a_0 plus all the previous inputs multiplied by their coefficients.

For example, a filter Transfer Function with poles and zeros is given in Equation 7.24.

$$\frac{z(z-0.8)(z+1)}{(z-0.5+j.7)(z-0.5-j0.7)(z-0.8)} \qquad (7.24)$$

The Scilab instructions to convert the factored form of 7.24 follow:

```
>> w=0.5+0.7*%i
w =
    0.5 + 0.7i
>> a=poly([w conj(w) 0.8],"z","r")
a =

   -0.592 +1.54z  -  1.54z2 + 1.8z3


>> b=poly([0 0.8 -1],"z","r")
b =

   -0.8z +0.2z2 +z3
```

Rewriting Equation 7.24 as a rational polynomial of variable z gives us Equation 7.25.

$$H(z) = \frac{Y(z)}{X(z)} = \frac{z^3 - 0.2z^2 - 0.8z}{z^3 - 1.8z^2 + 1.54z - 0.592} \qquad (7.25)$$

Rewriting the Equation 7.25 as the frequency response multiplied by the Transformed input value, we get Equation 7.26.

$$z^3 Y(z) - 1.8z^2 Y(z) + 1.54z Y(z) - 0.592 Y(z) = z^3 X(z) - 0.2z^2 X(z) - 0.8z X(z)) \quad (7.26)$$

The difference equation from Equation 7.26 is shown in Equation 7.27.

$$y[n+3] - 1.8y[n+2] + 1.54y[n+1] + 0.592y[n] = x[n+3] - 0.2x[n+2] - 0.8x[n+1] \; (7.27)$$

Digital Filters

There is just one problem with Equation 7.27: the solution requires values from the future. If n is the current sample, then $[n + 3]$ is the third sample in the future. Systems that depend upon values from the future are called *noncausal*, and it is not possible to implement them in real time. But if your system is linear and time invariant, the impulse response in the future is the same as impulse response in the past. Equation 7.29 could be shifted to the values from the past by simply subtracting 3 from each sample time. The sample time $[n + 3]$ becomes $[n]$ and $[n + 2]$ becomes $[n - 1]$, etc.; this makes the system a *causal* system so it may be implemented as a solution in real time. Thus, Equation 7.27 is equivalent to:

$$y[n] - 1.8y[n-1] + 1.54y[n-2] + 0.592y[n-3] = x[n] - 0.2x[n-1] - 0.8x[n-2]$$

Simplifying the previous equation as a function of variable $y[n]$, we obtain the desired convolution equation suitable for digital computer implementation, as shown in Equation 7.28.

$$y[n] = x[n] - 0.2x[n-1] - 0.8x[n-2] + 1.8y[n-1] - 1.54y[n-2] - 0.592y[n-3] \quad (7.28)$$

Equation 7.28 is the desired filter response corresponding to the poles and zeros from the Transfer Function of Equation 7.25. A recursive algorithm may be implemented to realize the output, but as you can see, it becomes difficult to compute the rational polynomial as the degrees of the terms are increased. There is a simpler way of dividing the polynomials into sections, as we'll discuss next.

Cascading Transfer Functions

If you think the number of the coefficients to multiply in one equation is overwhelming, you may reduce the complexity by dividing up the pole and zeros into sections and feeding the output of one to the input of the other. A real pole and real zero is a first order section, and a conjugate pole and zero is a second order section. Equation 7.25 may be reduced to the following sections:

Section 1:

$$\frac{U(z)}{X(z)} = \frac{z}{(z-0.8)}$$

$$u[n] = x[n] + 0.8u[n-1]$$

Section 2:

$$\frac{V(z)}{U(z)} = \frac{(z-0.8)}{(z-0.5+j.7)(z-0.5-j0.7)}$$

$$v[n] = u[n-2] - 0.8u[n-1] + v[n-1] + 0.74v[n-1]$$

Section 3:

$$\frac{Y(z)}{V(z)} = (z+1)$$

$$y[n] = v[n] + v[n-1]$$

Based on the techniques of poles and zeros mentioned earlier, we are ready to implement difference equations of some simple digital filters. The test program "filterTest.pl" described in Appendix C could be used to verify the outcome of the filters. The program generates a series of input values ($x[0], x[1]...x[n]$) corresponding to the test frequencies and feeds the values into the convolution equation of the filters and generates the corresponding output $y[0], y[1]...y[n]$.

Low-Pass Filter Realization

The Transfer Function of the low-pass filter of Example 7.1 is given as

$$\frac{Y_z}{X_z} = \frac{z}{z-r}$$

$$zY_z - rY_z = zX_z$$

The corresponding difference equation in the causal form is shown in Equation 7.29.

$$y[n] = x[n] + ry[n-1] \qquad (7.29)$$

Substituting the value of $r = 0.48$ and solving Equation 7.29 with an iterative algorithm, we realize a low-pass filter corresponding to Example 7.1. The test program "filterTest.pl" in Appendix C could be used to view the outcome of the filter for a given set of frequencies.

High-Pass Filter Realization

The Transfer Function of the high-pass filter of Example 7.2 is given as

$$\frac{Y_z}{X_z} = \frac{z}{z+r}$$

$$zY_z + rY_z = zX_z$$

The corresponding difference equation in the causal form is shown in Equation 7.30.

$$y[n] = x[n] - r*y[n-1] \qquad (7.30)$$

Substituting the value of $r = 0.48$ and solving Equation 7.33 with an iterative algorithm, we realize a high-pass filter corresponding to Example 7.2.

EXAMPLE 7.6

A data acquisition system has a sampling rate of 1024 samples per second. Formulate the difference equation that would suppress all frequencies beyond 20 Hz.

The Nyquist frequency:

$$1024/2 = 512\,Hz$$

The pole frequency as a fraction of the Nyquist frequency:

$20/512 = 0.039$ of the Nyquist frequency.

The pole distance:

$$r = (1 - 0.039) = 0.966$$

The maximum value of the peak is at 0°:

$$\theta = 0$$

The cutoff frequency vector:

$$re^{j0} = 0.966$$

The Transfer Function:

$$H(z) = \frac{Y(z)}{X(z)} = \frac{z}{z - 0.966} = \frac{1}{1 - 0.966z^{-1}}$$

The frequency domain multiplication:

$$Y(z) - 0.966 Y(z) z^{-1} = X(z)$$

The convolution in time domain is shown in Equation 7.31.

$$y[n] = x[n] + 0.966 y[n-1] \tag{7.31}$$

The time series output may be generated using the test program "filterTest.pl." Enter 0.966 for Pole0 magnitude and enter 0 for the Pole0 phase.

The Band-Pass and Band-Stop Filter Realization

The Transfer Function corresponding to the band-pass filter of Example 7.3 is,

$$H(z) = \frac{z^2}{z^2 - 2rz\cos\theta + r^2} = \frac{1}{1 - 2r\cos\theta z^{-1} + r^2 z^{-2}}$$

Substituting the required value of $\theta = 293\pi$ and $r = 0.995$ into the Transfer Function, we get Equation 7.32.

$$Y_z - 1.20 z Y_z + 0.99 Y_z z^{-2} = X_z \tag{7.32}$$

The causal form of the Equation 7.35 is shown in Equation 7.33.

$$y[n] = x[n] + 1.20 y[n-1] - 0.99 y[n-2] \tag{7.33}$$

Figure 7.13 displays outcome of the notch filter using the test program "filterTest.pl."

You can design simple filters by selecting poles and zero locations and viewing the response, but a precision filter requires other properties, such as a steeper roll-off of the pass-band region, etc. Ideally, one would like to establish a cutoff frequency and expect a filter to preserve the pass-band range while sharply rejecting the stop band. But such filters are not physically realizable, as they require infinite computations. Still, design methods are available to improve upon the roll-off rate and achieve a good enough response that closely matches the ideal response.

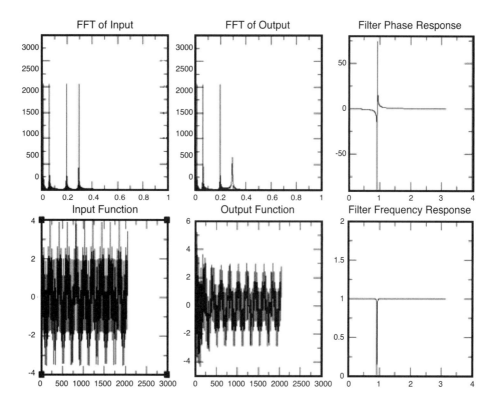

FIGURE 7.13 The output of the filterTest.pl program showing the notch filter. a) The sinusoidal input function. b) The FFT of the input. c) The filter output. d) FFT of the filter output. e) The frequency response. f) The phase response of the Transfer Function of the notch filter.

IMPROVING FILTER RESPONSE

One of the most important considerations in any filter design is to see how well the filter suppresses the frequencies in the roll-off region. We would like to see an action like a switch, all frequencies before the cutoff point appear in the system and the rest disappear. But if it is not possible, preferably, have as narrow a region as possible. There is a trade-off between the flat response and sharp cutoff frequency. Different methods enhance different aspects of the filter characteristics. In this chapter, we will discuss Butterworth design (which we mentioned in an earlier chapter) from the digital filter's point of view. We will also discuss Chebyshev's polynomials as an alternate way to improve the filter response.

BUTTERWORTH FILTERS

As explained in Chapter 6, "Filter Design," the Butterworth filters put extra weight by sharply increasing the pole magnitude at the cutoff point. It is like having multiple pole vectors at the same cutoff point.

$$|H(s)|^2 = |H(s)H(-s)| = \frac{1}{1+(s/j\omega_c)^{2N}}$$

The poles are the roots of the equation, shown in Equation 7.34.

$$s = (-1)^{1/2N} j\omega_c$$

Where $(-1)^{1/2N} (e^{j\pi(2k-1)})^{1/2N}$. For $k = 0, 1, 2..2N-1$
Therefore, each individual pole may be specified as

$$s_k = j\omega_c e^{j\pi(2k-1)/2N}$$
$$s_k = e^{j\pi/2} \cdot \omega_c e^{j\pi(2k-1)/2N} = \omega_c e^{j\pi(2k+N-1)/2N} \quad (7.34)$$

Equation 7.34 indicates that there are N numbers of pole pairs (see Chapter 6 for a detailed explanation). Half the poles belong to the right side of the S plane and are not realizable. The realizable poles on the left side of the S plane for even values of N are,

$N = 2$
$$\frac{1}{(z - e^{j\pi 3/4})(z - e^{-j\pi 5/4})}$$

$N = 4$
$$\frac{1}{(z - e^{j\pi 5/8})(z - e^{-j\pi 7/8})(z - e^{j\pi 9/8})(z - e^{-j\pi 11/8})}$$

Butterworth filters may be realized by cascading several single pole filters, where the output of one is the input to others.

Matlab and Scilab can simplify the design process as they are provided with built-in functions to compute the coefficients.

The following examples use Matlab syntax to generate the coefficients and display the frequency response of the filters. The numerator coefficients are in the b vector and denominator coefficients are in the a vector.

EXAMPLE 7.7

Design a sixth order low-pass Butterworth filter having the cutoff frequency of 100 Hz for a sampling rate of 500 Hz.

```
>> [b,a] = butter(6,100/250);
>> freqz(b,a,256,500);
```

See Figure 7.14 for the frequency and phase response.

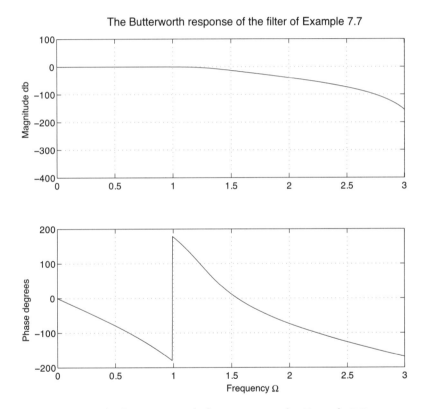

FIGURE 7.14 The frequency and phase response for Example 7.7.

EXAMPLE 7.8

Design a ninth order high-pass Butterworth filter having the cutoff frequency of 300 Hz for a sampling rate of 1000 Hz.
 The cutoff point at Nyquist frequency is 3/5=0.6.

```
>> [b,a] = butter(9,300/250,'high');
>> freqz(b,a,256,1000);
```

See Figure 7.15 for the frequency and phase response.

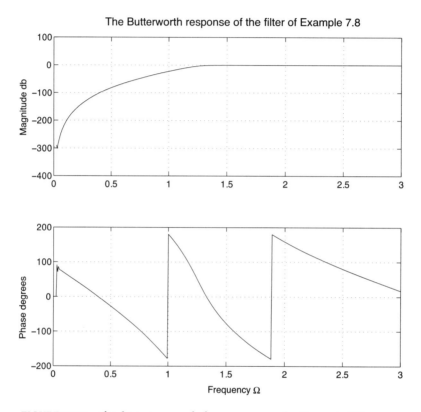

FIGURE 7.15 The frequency and phase response for Example 7.8.

EXAMPLE 7.9

Design a tenth order band-pass Butterworth filter having the bandwidth of 100 Hz centering on 150 Hz for a sampling rate of 1000 Hz.

```
>> W = [100 200]/500;
>> [b,a] = butter(10, W);
>> freqz(b,a,256,1000);
```

See Figure 7.16 for the frequency and phase response.

Digital Filters   253

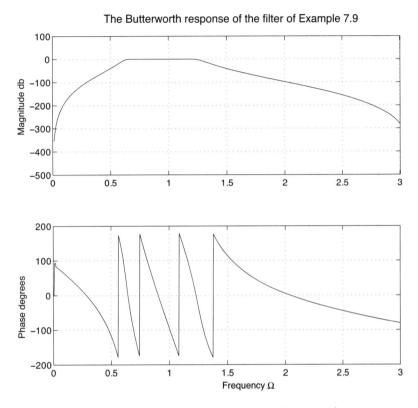

FIGURE 7.16   The frequency and phase response for Example 7.9.

EXAMPLE 7.10

Design a tenth order band-stop Butterworth filter having the bandwidth of 100 Hz centering on 150 Hz for a sampling rate of 1000 Hz.

```
>> W = [100 200]/500;
>> [b,a] = butter(10, W,'bandstop');
>> freqz(b,a,256,1000);
```

See Figure 7.17 for the frequency and phase response.

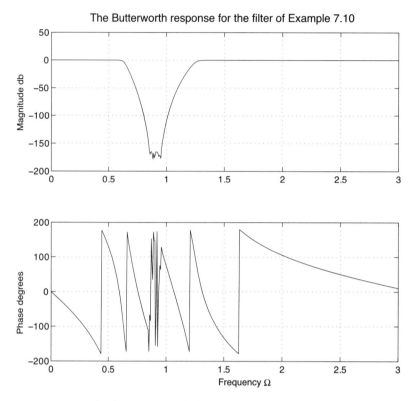

FIGURE 7.17 The frequency and phase response for Example 7.10.

CHEBYSHEV FILTERS

Chebyshev filters provide a polynomial fit (as a weighted function) to a given Transfer Function, but there are consequences in the form of ripples in the pass-band or the stop-band region.

The Chebyshev's series is defined in Equation 7.35.

$$T_n(\omega) = \cos(n \arccos \omega) \qquad (7.35)$$

Figure 7.18 is the plot of the series with increasing n. Notice that the points on the x-axis are much closer at the end; this unequal spacing tends to put more weight at the end to improve upon the response near the edge.

The first few terms of the series are expanded as,

$$T_0(\omega) = 1$$
$$T_1(\omega) = \omega$$
$$T_2(\omega) = 2\omega^2 - 1$$
$$T_3(\omega) = 4\omega^3 - 3\omega$$
$$T_4(\omega) = 8\omega^4 - 8\omega^2 + 1$$
$$T_5(\omega) = 16\omega^5 - 20\omega^3 + 5\omega$$
$$T_6(\omega) = 32\omega^6 - 48\omega^4 + 18\omega^2 - 1$$
$$T_7(\omega) = 64\omega^7 - 112\omega^5 + 56\omega^3 - 7\omega$$
$$\vdots$$
$$T_{n+1}(\omega) - 2\omega T_n(\omega) + T_{n-1}(\omega) = 0$$

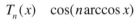

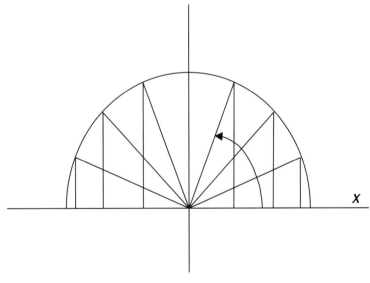

FIGURE 7.18 The magnitude of Chebyshev series on the x-axis showing increase in width.

The function has the property that for $x < 1$ it is cosine function but for $x > 1$ it becomes a hyperbolic function. If we normalize w with the cutoff frequency wc, then for $w/wc < 1$ we have cosine function and for $w/wc > 1$ we have hyperbolic function, producing a steep drop in the magnitude of the Transfer Function at the cutoff point of $w/wc = 1$, which is also the transition point from pass band to stop band.

$$T_N(\omega) = \begin{vmatrix} \cos(N \cos^{-1} \omega) & |\omega| \leq 1 \\ \cosh(N \cos^{-1} \omega) & |\omega| > 1 \end{vmatrix}$$

The Transfer Function generated by the Chebyshev polynomials is given as in Equation 7.36.

$$|H(\omega)|^2 = \frac{1}{1 + \varepsilon^2 [T_n(\omega/\omega_c)]^2} \tag{7.36}$$

The Chebyshev filters do enable us to achieve a sharper transition (steeper roll-off), but on the expense of some distortion (ripples) in the pass-band or stop-band region.

The ω_c in Equation 7.36 is the cutoff frequency and as long as $|\omega_c| \ll |\omega|$, the term $H(\omega)$ has the following range:

$$\frac{1}{1+\varepsilon^2} \leq |H(\omega)| \leq 1$$

And at $\omega_c = \omega$, which is also the edge, we have $T_n(1) = 1$ indicating the gain is 1 at the cutoff point, which meets our criteria.

The ripples produced by the filter in the pass-band region are bounded by the interval $\frac{1}{1+\varepsilon^2} \leq \frac{1}{1+\varepsilon^2[T_n(\omega/\omega_c)]} \leq 1$.

The condition at the edge of the stop-band where $\omega_s = \omega$ is

$$\frac{1}{1 + \varepsilon^2 T_n^2(\omega_s/\omega_c)} = \frac{1}{1 + A^2}$$

The filter order N may be calculated with the specifications of the magnitudes at the two critical frequencies, i.e., ω_c and ω_s as follows,

$$T_n(\omega_s/\omega_c) = \sqrt{\frac{A}{\varepsilon}}$$

Matlab software has built-in functions to compute the filter order; see Appendix A for details.

The following examples use Matlab syntax to generate the coefficients and display the frequency response of the filters. The numerator coefficients are in the *b* vector and denominator coefficients are in the *a* vector. The ripple parameters are specified in db.

EXAMPLE 7.11

Design a ninth order low-pass Chebyshev filter with 0.6 db ripples in pass-band having a cutoff frequency of 300 Hz for a sampling rate of 1000 Hz.

The corresponding Nyquist frequency is 1000/2 = 500.

```
>> [b,a] = cheby1(9,0.6,300/500);
>> freqz(b,a,256,1000);
```

See Figure 7.19 for the frequency and phase response.

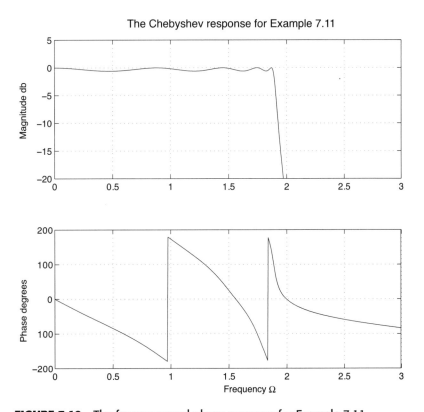

FIGURE 7.19 The frequency and phase response for Example 7.11.

EXAMPLE 7.12

Design a ninth order high-pass Chebyshev filter with 0.6 db ripples in pass-band having a cutoff frequency of 300 Hz for a sampling rate of 1000 Hz.

```
>> [b,a] = cheby1(9,0.6,300/500,'high');
>> freqz(b,a,256,1000);
```

See Figure 7.20 for the frequency and phase response.

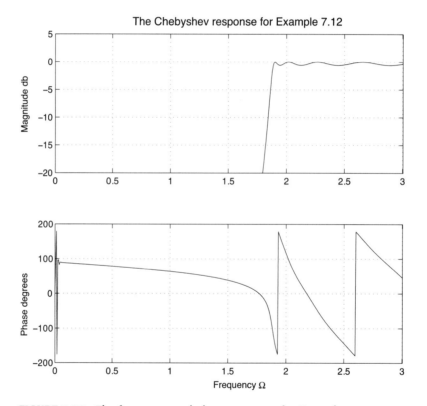

FIGURE 7.20 The frequency and phase response for Example 7.12.

EXAMPLE 7.13

Design a ninth order band-pass Chebyshev filter with 0.6 db ripples with pass-band of 100 to 200 Hz, for a sampling rate of 1000 Hz.

```
>> W = [100 200]/500;
>> [b,a] = cheby1(9,0.6,W);
>> freqz(b,a,256,1000);
```

See Figure 7.21 for the frequency and phase response.

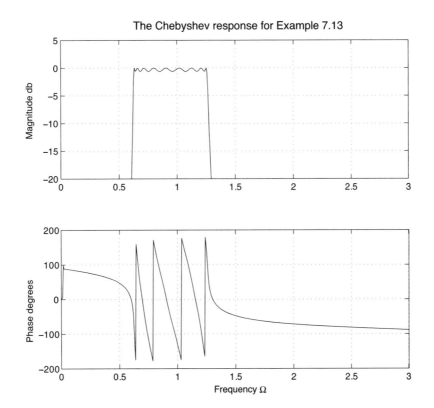

FIGURE 7.21 The frequency and phase response for Example 7.13.

EXAMPLE 7.14

Design a ninth order band-stop Chebyshev filter with 0.6 db ripples with band-stop range of 100 to 200 Hz. Sampling rate is 1000 Hz.

```
>> W = [100 200]/500;
>> [b,a] = cheby1(9,0.6,W);
>> freqz(b,a,256,1000);
```

See Figure 7.22 for the frequency and phase response.

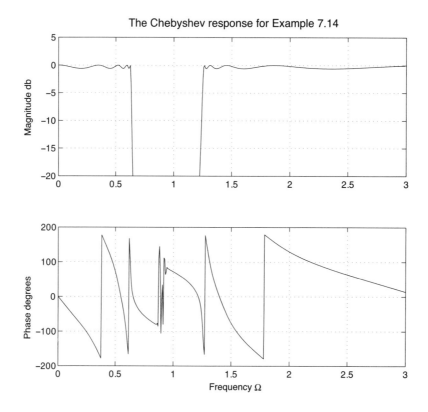

FIGURE 7.22 The frequency and phase response for Example 7.14.

Chebyshev Filters with Ripples in Stop Band

Chebyshev filters could be designed with ripples in stop band. The Matlab function for ripples in stop band is cheby2(...). The following examples use Matlab syntax to generate the frequency and phase response of the filters.

EXAMPLE 7.15

Design a ninth order band-stop Chebyshev filter with 0.6 db ripples in stop-band region with band-stop range of 100 to 200 Hz. Sampling rate is 1000 Hz.

```
>> W = [100 200]/500;
>> [b,a] = cheby2(9,0.6,W);
>> freqz(b,a,256,1000);
```

See Figure 7.23 for the frequency and phase response.

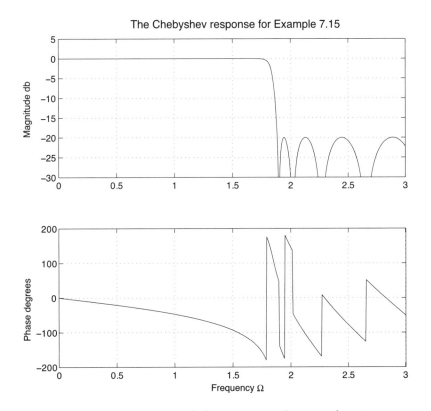

FIGURE 7.23   The frequency and phase response for Example 7.15.

ELLIPTIC FILTERS

Elliptic filters offer steeper roll-off than Butterworth or Chebyshev filters, but they produce ripples in pass-band as well as stop-band regions. The Matlab function for elliptic filter is ellip(...). You may specify attenuation in the stop-band region as a parameter to the function.

EXAMPLE 7.16

Design a sixth order band-stop elliptic filter with 3 db ripples in pass-band and 50db drop in pass-band region with band-stop range of 100 to 200 Hz. Sampling rate is 1000 Hz.

```
>> W = [100 200]/500;
>> [b,a] = ellip(6,3,50,W);
>> freqz(b,a,256,1000);
```

See Figure 7.24 for the frequency and phase response.

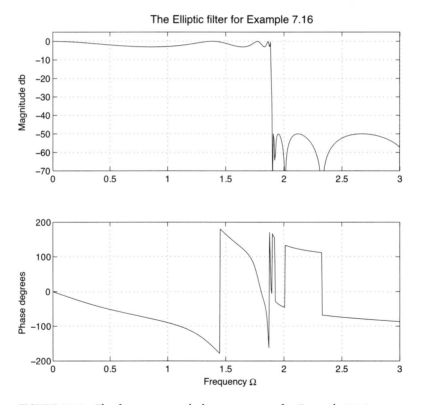

FIGURE 7.24 The frequency and phase response for Example 7.16.

THE TRANSITION FROM ANALOG TO DIGITAL FILTERS

Transforming an analog filter into digital form in the z-Transform is basically an operation of frequency warping, because the response of the analog filters that theoretically extends to infinity should be mapped to a function that extends to Nyquist frequency or one cycle of π radians per sample only.

The frequency variable ω_a of the analog filters' Transfer Function, which is a function in sine and cosine, is replaced with the tangent function in Equation 7.37 (using half angle formula of trigonometry), which has digital discrete frequency ω_d of radians per sample.

$$\omega_a = \frac{2}{\Delta t} \tan \frac{\omega_d \Delta t}{2} \tag{7.37}$$

Let's take a simple analog low-pass RC filter of a single pole with the Transfer Function in Equation 7.38.

$$H(\omega_a) = \frac{A}{\omega_a + p_i} \tag{7.38}$$

Substitute the value of ω_a as given in Equation 7.37, and you get Equation 7.39.

$$H(\omega_a) = \frac{A}{j(\frac{2}{\Delta t} \tan \frac{\omega_d \Delta t}{2}) + p_i} \tag{7.39}$$

The response of the analog filter and the response from the transform of Equation 7.39 is identical up to the cutoff frequency, but then the analog response approaches 0 only at frequency $\omega_a \to \infty$, whereas the digital filter response approaches 0 at the Nyquist frequency. Beyond the Nyquist point, the digital response repeats the pattern for the next increment of π radians,

Replacing the analog frequency variable with its equivalent digital frequency variable function serves our purpose of converting an analog filter into its equivalent digital filter.

We can rewrite Equation 7.39 into the familiar exponent form shown in Equation 7.40.

$$\frac{2}{\Delta t} \tan \frac{\omega_d \Delta t}{2} = \frac{2}{\Delta t} \frac{1 - e^{-\omega_d \Delta t}}{1 + e^{-\omega_d \Delta t}} \tag{7.40}$$

Substituting the standard notation of $z = e^{\omega_d \Delta t}$ into Equation 7.41,

$$\omega_a = \frac{2}{\Delta t} \frac{1-z^{-1}}{1+z^{-1}} \qquad (7.41)$$

The relationship in Equation 7.41 is known as a *bilinear z-Transform* that relates an analog frequency to its digital form.

There is a small amount of distortion when we transform analog frequencies into digital frequencies using the bilinear z-Transform of Equation 7.40. The distortion may be corrected by first obtaining the corresponding analog frequencies that matched the Transfer Function and then applying the transformation using a bilinear z-Transform. The steps are as follows:

1. Specify the desired digital filter specifications.
2. Obtain the corresponding analog frequencies using Equation 7.40.
3. Design the analog prototype filter $H(s)$.
4. Convert the analog prototype into the target analog filter using a frequency-frequency transform.
5. Design a digital filter by substituting the analog frequency variable with the z-Transform of Equation 7.41.

EXAMPLE 7.17

Let's analyze the analog *RC* filter response by replacing it with the bi-linear z-Transform of Equation 7.41; shown in Equation 7.42.

$$H(\omega_a) = \frac{\omega_a}{\omega_a + p}$$

$$H(\omega_a) = \frac{(\omega_a \Delta t/2)(1+z^{-1})}{((\omega_a/2)+1)+(\omega_a \Delta t/2 - 1)z^{-1}} \qquad (7.42)$$

Expanding Equation 7.42 in terms of z,

$$H(z) = \frac{(\omega_a \Delta t/2)+(\omega_a \Delta t/2 z^{-1})}{((\omega_a/2)+1)+(\omega_a \Delta t/2 - 1)z^{-1}}$$

And we obtain the difference equation as shown in Equation 7.43.

$$y(n) = \omega_a \Delta t/2\, x[n] + \omega_a \Delta t/2\, x[n-1] - (\omega_a/2 + 1)y[n] - (1 + \omega_a \Delta t/2)y[n-1] \quad (7.43)$$

EXAMPLE 7.18

What is the equivalent cutoff analog frequency, if a digital filter is designed with the sampling rate of 1K samples per second and a cutoff frequency of 700π radians per sample?

From the sampling rate, we compute the sampling period:

$$\Delta t = 2\pi/1000 = .002\pi \text{ rad s}^{-1}$$

The sampling frequency is

$$\omega_s = 2\pi \times 10^3 \text{ rad s}^{-1}$$

The digital filter cutoff point:

$$\omega_{dc} \Delta t = (0.7 \times 10^3).002\pi = 0.2\pi$$

The equivalent analog filter cutoff is,

$$\omega_{ac} \Delta t/2 = \tan(\omega_{dc} \Delta t/2) = \tan(0.2\pi) = 0.3$$

EXAMPLE 7.19

Design a digital filter with no ripples in the pass-band or stop-band region. The sampling rate is 1 K samples per second with a cutoff frequency of 1 KHz. The filter must reduce frequencies beyond 1.5 KHz to more than 20 dB.

Choice is a Butterworth filter, since Chebyshev and elliptical filters have ripples in the pass-band region.

The frequency in radian per sample:

$$\omega_{d1}\Delta t = 2\pi \times 10^3 \times (1/10^3) = 2\pi \text{ rad s}^{-1}$$
$$\omega_{d2}\Delta t = 2\pi \times 1.5 \times 10^3 \times (1/10^3) = 3\pi \text{ rad s}^{-1}$$

The corresponding analog frequencies are,

$$\omega_{a1} = \tan(\omega_{d1}\Delta t/22) = \tan(0.1\pi) = 0.325$$
$$\omega_{a2} = \tan(\omega_{d2}\Delta t/22) = \tan(0.2\pi) = 0.726$$

The magnitude of the Butterworth nth order filter from Equation 6.1 is given by,

$$|H(\omega)|^2 = \frac{1}{1+(\omega/\omega_c)^{2n}}$$

Where $\omega_c = \omega_{a1} = 0.325$. The order n to be computed from the condition of 20 dB down at $\omega_{a2} = 0.726$

$$1 + (0.726/0.325)2^n = 20$$

$$n = 1.83$$

We take the next increment

$$n = 2$$

The poles of Butterworth s_1 and s_2 are shown below; see Equation 7.43.

$$s_1 = s_2 = 0.325(-0.707 \pm j0.707) = -0.23 \pm j0.23 \text{ And no zeros}$$

$$H(s) = \frac{s_1 s_2}{(s+s_1)(s+s_2)} = \frac{0.1058}{s^2 + 0.46s + 0.1058} \quad (7.43)$$

Replacing s with bilinear z-Transform $s = \frac{1-z^{-1}}{1+z^{-1}}$ in Equation 7.44,

$$H(z) = \frac{0.1058}{\left(\frac{1-z^{-1}}{1+z^{-1}}\right)^2 + 0.46\left(\frac{1-z^{-1}}{1+z^{-1}}\right) + 0.1058} \quad (7.44)$$

Simplifying Equation 7.44,

$$H(z) = 0.0676 \frac{1 + 2z^{-1} + z^{-2}}{1 - 1.141z^{-1} + 0.413z^{-2}}$$

The corresponding difference equation

$$y(k) - 1.141y(k-1) + 0.413y(k-2) = 0.0676x(k) + 0.135x(k-1) + 0.676x(k-2)$$

That can be computed with an iterative computer algorithm.

The frequency response is given by

$$H(\omega) = 0.0676 \frac{1 + 2e^{-j\omega\Delta t} + e^{-j2\omega\Delta t}}{1 - 1.141e^{-j\omega\Delta t} + 0.413e^{-j2\omega\Delta t}}$$

The Matlab function for the bilinear transform is `bilinear(...)`. It accepts frequency domain parameters and returns the mapped version of the variables for the digital domain. See Appendix A for examples using Matlab syntax.

SUMMARY

In this chapter, we used poles and zeros to formulate the factored form Transfer Function corresponding to the filter frequency response. The filter output was realized through implementing convolution as the solution of the difference equation, whose coefficients were obtained from the filter Transfer Function. The frequency and phase response was plotted using graphical user interface test software. The program also generated the coefficient values corresponding to the difference equation that formed the filter Transfer Function. The user could enter a choice of test patterns and sampling rate, and the program would generate filter output for one complete cycle.

8 The FIR Filters

In This Chapter

- Ideal Filters
- FIR Filter Design
- Windowing

Digital filters are classified as IIR, or Infinite Impulse Response, filters and FIR, or Finite Impulse Response filters. In IIR filters, the previous out- and input are fed back to the system after multiplying with a complex quantity, whereas in the FIR filters, only the previous input is used in the feedback part, thus providing a linear phase response. We have discussed the IIR filters in the previous chapter and designed a few, including the high-pass, low-pass, band-pass, and band-stop filters. In all, the computation time was fast, as it did not take much to calculate a few poles and zeros, but the phase response was generally poor. The initial transient response in IIR filters lingers on for a long time. The filters require values from the past output, and some parts of the initial values are always present in the current output, in theory at least. Though, in practice, the fraction grows less and less and eventually reaches 0, it might take a very long time before the floating-point arithmetic precision gives way to the accuracy of the calculated numbers. The linear phase response makes the FIR filters suitable candidates where the phase information has an important role to play.

We now discuss a different method of obtaining a filter Transfer Function that would simply rely upon the values from the past input but not the past outputs, as such; there is a uniform phase delay with no distortion. The linear phase is a definite plus, where phase delay is critical, and this is the Fourier Transform method of designing digital filters. The figures and the tables in this chapter were created with the help of Scilab and Matlab scripts, and listings of the instructions are provided in Appendixes A and B. It is recommended that the reader familiarize themselves with the instructions in the scripts and gain confidence in filter design method by seeing the response interactively. The starting point for the FIR filter design is the ideal frequency response.

IDEAL FILTERS

Figure 8.1 is a repeat of the ideal filter response that we have seen before. Notice the rectangular shape and the sharp cutoff point that marks the transition from the pass band to the stop band, there is practically no transition band, and the gain is either 0 or 1. This is an ideal filter, and if we could somehow achieve it, it would solve all our filtering problems, but in reality (as you will see later), we cannot achieve such perfection. We can only try to get closer. In FIR filters, we seek the function that closely matches the ideal response.

Let's work backward and think; if there was a frequency response close to the ideal response, there must have been a time-based impulse response that generated the ideal frequency response. The Fourier Transform of the ideal response would have impulse response as its coefficients, and this is the basis of the Fourier Transform method of designing digital filters. Once we know the impulse response, it is easier to formulate a filter by performing convolution of the input samples with the impulse response, just as we did in the previous section of the IIR filter design.

The objective is to find the coefficients of the Fourier series that generated the ideal response (or close to it) and that Fourier series is the desired filter Transfer Function.

Let's choose a low-pass filter as shown in Figure 8.2. (As usual, we will be denoting the discrete sampling frequency with the symbol Ω.) Ignoring for a moment the negative frequencies, such an ideal frequency response has a well-defined period $-\pi$ to $+\pi$ with the pass-band region from $-\Omega$ to $+\Omega$ (where Ω is the normalized sampling frequency of radians per sample), centered on 0. This is a perfect low-pass filter with the cutoff sampling frequency of $+\Omega$. The values within the pass-band region are 1 and outside this region, they are 0, thus forming a function with the values as shown in the following series,

0,0,0,..0,0,0,1,1,1...1,1,1,0,0,0..0,0,0

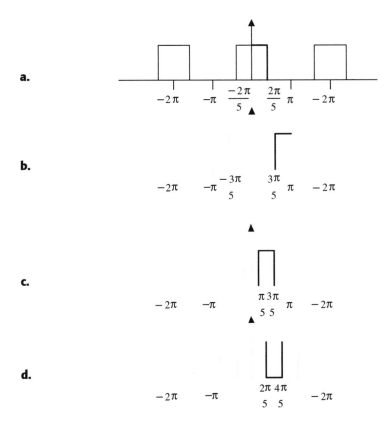

FIGURE 8.1 The ideal frequency response. a) Low-pass filter. b) High-pass filter. c) Band-pass filter. d) Band-stop filter.

What function has such a rectangular shape? To answer this question, let's revisit the Fourier series and produce the coefficients that form the output.

The Fourier Series Representation of the Ideal Frequency Response

A discontinuous function such as the one in Equation 8.2 may be expressed in terms of its Fourier series. The function itself is a periodic function, and according to the definition, a periodic function $f(x)$ may be represented as an infinite series of sine and cosine, whose amplitudes (or coefficients) c_n may be computed from the function itself, as shown in Equations 8.1 and 8.2. (See Chapter 3, "Fourier Transform," for a discussion on computing the Fourier coefficients.)

$$f(t) = \sum_{n=-\infty}^{\infty} c_n e^{j\omega t} \qquad (8.1)$$

The coefficients:

$$c_n = \frac{1}{2\pi} \int_0^\pi f(t)e^{-j\omega x} dt \qquad (8.2)$$

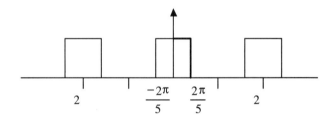

FIGURE 8.2 The ideal low-pass filter with the cutoff frequency $-\Omega = \frac{-2\pi}{5}$ to $+\Omega = \frac{2\pi}{5}$.

In digital filters, we have a reverse situation: the frequency response $H(\Omega)$ is a periodic function, and the coefficients are time-based. The discrete time Fourier series of the ideal filter Transfer Function of Figure 8.2 is given in Equation 8.3.

$$H(\Omega) = \sum_{n=-\infty}^{\infty} h_n e^{-j\Omega n} \qquad (8.3)$$

The coefficients h_n are obtained using the inverse discrete time Fourier Transform, shown in Equation 8.4.

$$h(n) = \frac{1}{\Omega_{-c} - \Omega_c} \int_{\Omega_{-c}}^{\Omega_c} H(\Omega)e^{j\Omega n} d\Omega \qquad (8.4)$$

Note that the periodic response function $h(n)$ of Equation 8.4 has a period of $-\pi$ to $+\pi$, but the pass-band region from $\Omega_{-c} = -2\pi/5$ to $\Omega_c = 2\pi/5$ (example form Figure 8.1) is the only nonzero part, thus, the integral is performed over the pass-band region only. In our case $\Omega_c = -\Omega_c$, substituting the value of the Transfer Function (between Ω_{-c} to Ω_c) and integrating to obtain the expression for the impulse response $h(n)$ we get,

$$h(n) = \frac{1}{2\pi} \int_{-\Omega_c}^{\Omega_c} 1 \bullet e^{jn\Omega} d\Omega = \frac{1}{2\pi} \frac{e^{jn\Omega}}{jn} \Big]_{-\Omega_c}^{\Omega_c}$$

$$h(n) = \frac{\Omega_c}{\pi} \times \frac{\sin(n\Omega_c)}{n\Omega_c}$$

Substituting the value of h_n in Equation 8.3, we get the Transfer Function $H(\Omega)$ as shown in Equation 8.5, essentially, an inverse Fourier Transform of a rectangular pulse.

$$H(\Omega) = \sum_{n=-\infty}^{\infty} (\frac{\Omega_c}{\pi} \times \frac{\sin(n\Omega_c)}{n\Omega_c}) e^{-j\Omega n} \qquad (8.5)$$

The coefficients $h(n)$ of Equation 8.5 are equivalent of the time-based impulse response of the ideal frequency response. According to Fourier, we need infinite numbers of these coefficients to truly represent the given periodic function.

From a practical standpoint, the number of coefficients is a matter of preference; since a filter with infinite coefficients is inconceivable, we limit ourselves to a finite number N. We will see the consequences of choosing a higher or lower value later in the section, but next, we discuss its magnitude response obtained from the filter Transfer Function.

The Filter Transfer Function

A Transfer Function of a filter is a function of frequency and it is the Fourier Transform of the filter's impulse response. It is also the ratio of the Transform of the output over the Transform of the input. A plot of the Transfer Function versus frequency variable Ω shows how the output magnitude varies with the different input frequencies. Taking the Fourier Transform of the impulse response of the ideal filter, we describe the Transfer Function as,

$$H(\Omega) = \frac{Y_z}{X_z} = \sum_{n=-\infty}^{\infty} h_n e^{-jn\Omega}$$

The ideal frequency response requires infinite summation, but suppose we truncate the series to a finite $2N + 1$ number of terms. The Transfer Function evaluated with these reduced numbers is given in Equation 8.6.

$$H(\Omega) = \sum_{n=-N}^{N} h_n e^{-j\Omega n}$$

$$H(\Omega) = h_{-N} e^{j\Omega N} + h_{-(N-1)} e^{j\Omega(N-1)} \cdots h_0 e^0 \cdots h_{(N-1)} e^{-j\Omega(N-1)} + h_N e^{-j\Omega N} \qquad (8.6)$$

Equation 8.6 is simply a polynomial in h_n evaluated on the unit circle. (See Appendix B, "Scilab Tutorial," for the Scilab instructions to perform the evaluation.) But we could also use the following derivative and obtain a close form expression of the filter Transfer Function.

Substituting the equality,

$$h_{-n} = h_n$$

$$e^{-j\Omega n} + e^{j\Omega n} = 2\cos(\Omega n)$$

We obtain the following equation for the Transfer Function (Equation 8.7).

$$H(\Omega) = h(0) + 2h_1 \cos\Omega + 2h_2 \cos 2\Omega + \cdots + 2h_{n-1}\cos(N-1)\Omega + 2h_N \cos N\Omega$$

$$H(\Omega) = h(0) + 2\sum_{k=1}^{N} h(n)\cos k\Omega \qquad (8.7)$$

There is no imaginary component in Equation 8.7 that means the filter has linear phase response, and the Transfer Function magnitude is simply,

$$|H(\Omega)| = h(0) + 2\sum_{k=1}^{N} h(n)\cos k\Omega$$

Substituting the values for h_n computed with Equation 8.7, we obtain an expression for the amplitude response.

With the knowledge of filter coefficients and the design of magnitude response, we are ready to implement FIR filters.

FIR FILTER DESIGN

We will be discussing the four different kinds of filters: the low-pass, high-pass, band-pass, and band-stop filters. The frequency response of each is depicted in Figure 8.1. Implementing a filter is simply a matter of finding the right impulse response and developing a convolution algorithm that convolves the input samples with the appropriate impulse response (similar to the IIR filter of Chapter 7, "Digital Filters"). Thus, the quest is to obtain the right impulse response. We begin with the implementation of a low-pass filter.

The Low-Pass FIR Filter

A filter is known by its frequency response and is characterized by its cutoff frequencies. From an implementation point of view, the number of coefficients or filter-taps matter as they affect the speed of execution. Though choosing a large number of filter coefficients brings the filter response close to the ideal response, it may slow down the execution. In our example filter, we chose 21 coefficients. The cutoff frequency is determined with respect to the sampling frequency. If the sampling rate is 10k samples per second, a cutoff frequency of 2.5kHz means $\Omega_c = 0.25\pi$.

As usual, we will perform convolution of the 21 points (of the impulse response) with the *kth* input samples $x[k]$, to produce the *kth* filter output $y[k]$, the reason for picking an odd number is to have symmetry on 0.

$$y[k] = \sum_{n=-10}^{10} h_n x[k-n]$$

The first step in filter design is to compute the impulse response corresponding to the frequency response. We have already developed an expression for the low-pass filter by taking the inverse Fourier Transform, as shown in the derivative of Equation 8.5. Ideally, we should evaluate the equation for $n = -\infty$ to $n = +\infty$, but for 21 coefficients, we would use $n = -10$ to $n = +10$. For $n \neq 0$, the evaluation is straightforward, but to avoid the divide by 0 error (in case of $n = 0$) L'Hopital's rule could be used (Equation 8.8).

$$h[0] = \frac{(d/dn)\Omega_c \sin(n\Omega_c)}{(d/dn)n\pi} = \frac{\Omega_c \cos(n\Omega_c)}{\pi} = \frac{\Omega_c}{\pi} \quad \text{for } n = 0$$

$$h[n] = \frac{\Omega_c}{\pi} \times \frac{\sin(n\Omega_c)}{n\Omega_c} \quad \text{for } n \neq 0 \quad (8.8)$$

The impulse response given in Equation 8.8 is of the form $\sin \pi x/\pi x$, commonly known as a *sinc* function. Rectangular functions of either time or frequency domain usually transform into sinc functions when transformed from one domain to another. Matlab and the public domain software Scilab have a built-in sinc function that you can use to compute the coefficients of the desired filter. The coefficients given in Table 8.1 were generated with the help of the Scilab program, and the listing is provided in Appendix B.

EXAMPLE 8.1

Design a low-pass filter with the cutoff frequency of 1/4th of the Nyquist frequency.

We merely have to find the filter coefficients corresponding to the cutoff frequency. It should be noted that the actual cutoff frequency is two times the value of the desired frequency. Substituting the value of $\Omega_c = 0.25\pi$ and evaluating Equation 8.1 for $n = -10$ to $n = +10$, we get 21 coefficients, as described in Table 8.1.

The following are Matlab instructions for generating filter coefficients:

```
>>0.25*sinc(0.25*(-10:10))
```

Table 8.1 describes the filter coefficients $h(n)$ for $n = -10$ to $n = +10$. See Appendix B for the Scilab instructions to generate the coefficients.

TABLE 8.1 Filter Coefficients

h(−10)=h(10)	.0318310
h(−9)=h(9)	.0250088
h(−8)=h(8)	0
h(−7)=h(7)	−.0321542
h(−6)=h(6)	−.0530516
h(−5)=h(5)	−.0450158
h(−4)=h(4)	0
h(−3)=h(3)	.0750264
h(−2)=h(2)	.1591549
h(−1)=h(1)	.2250791
h(0)	.25

Convolving the impulse response with the input samples produces the filter output shown in Equation 8.9.

$$y[k] = \sum_{n=-10}^{10} h_n x[k-n] \quad (8.9)$$

If $k = 0$ is the current instance, Equation 8.9 is expanded as,

$$y[0] = h_{-10}x[-10] + h_{-9}x[-9] \cdots h_0 x[0] \cdots + h_{10}x[10] + h_9 x[9]$$

You may recognize that the filter output of Equation 8.9 is noncausal, as it requires values from the future. One way to solve the problem is to shift the filter coefficients so that they start from 0, instead of −10, as shown in Figure 8.3. Being a linear and time-invariant system, the impulse response in the future is the same as impulse response in the past. Our modified causal filter is,

$$y[k] = \sum_{n=0}^{20} h_n x[k-n]$$

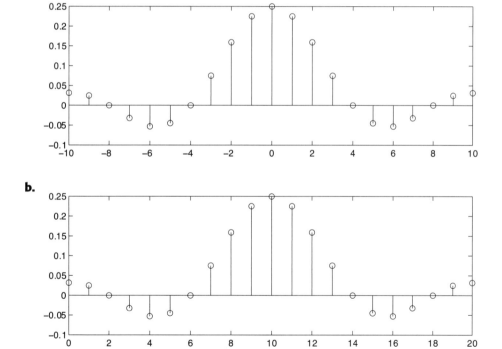

FIGURE 8.3 Impulse response of an ideal low-pass filter with cutoff frequency of 1/4th of the Nyquist. a) Noncausal. b) Causal.

Figure 8.4 is a plot of the frequency response of the low-pass filter of Example 8.1, together with the ideal frequency response for a comparison.

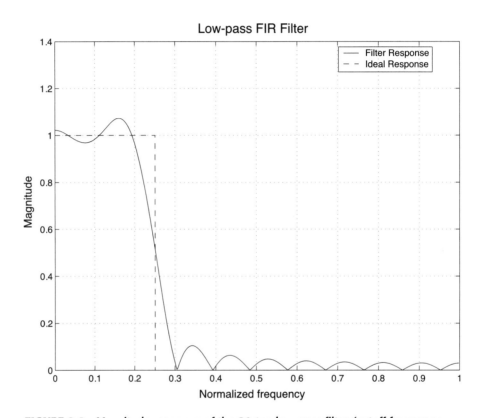

FIGURE 8.4 Magnitude response of the 21-tap low-pass filter (cutoff frequency $+\Omega = \frac{\pi}{4}$) compared to the ideal filter.

A logarithmic Bode plot of the frequency response of Example 8.1 is presented in Figure 8.5 to highlight the side lobe amplitude The normal plot shows the passband ripples overshoot of 9% (due to Gibb's phenomenon) near the transition band.

Decimation Filters

Sometimes we acquire signals with a far higher rate than our processing or storage can manage. It would be wise to pass the data through a low-pass filter before throwing away the unwanted, simply because there are higher frequency components that may produce an aliasing effect on the remaining data. Similarly, there are times

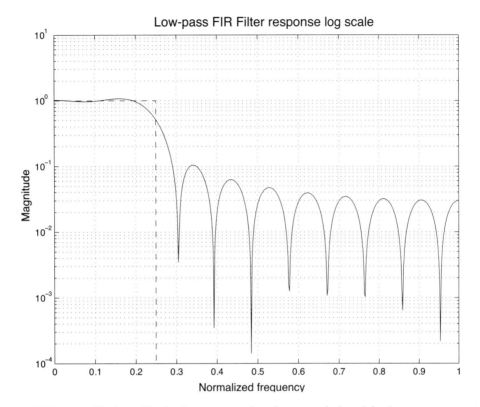

FIGURE 8.5 The logarithmic plot compared to the normal plot of the frequency response of the low-pass filter of Example 8.1, showing the ripple effect.

when data are acquired at a lower sampling rate and may need to be converted into fractional higher sampling rate. *Decimation* is the process of altering the sampling rate, and the *decimation factor* is the ratio of the input rate to the output rate.

The decimation ratio must be an integer since we cannot throw away part of the value. If there is a need for fractional reduction, interpolation must be performed before decimating the data. For example, to change the data recorded on a Digital Audio Tape (DAT format), which has a 48 kHz sampling rate, to Compact Disc (CD format), which has a 44.1 kHz sampling rate, find the integer conversion ratio p/q, up-sample with zero insertion by the factor of p, and filter through a low-pass filter with a cutoff frequency $\omega_c = 2\pi/q$, and then down sample (throw away data) by a factor of q. Matlab function upfirdn performs such decimation: up-sampling of the x data by the factor p, filtering by the filter coefficient h, and down-sampling by the factor q. Matlab can perform these steps with the following commands,

```
>> D=gcd(44100,48000)
>> P=44100/d;
>> Q=48000/d;
>> Y=upfirdn(x,h,p,q);
```

The decimation can efficiently be performed in multiple stages, so long as the decimation factor, M, is not a prime number. For example, to decimate by a factor of 15, you could decimate by 5, then decimate by 3. The more prime factors M has, the more choices you have. For example, you could decimate by a factor of 12 using:

- One stage: 12
- Two stages: 3 and 4 or 6 and 2
- Three stages: 2, 3, and 2

From a software implementation point of view, a decimating FIR is actually the same as a regular FIR, except that you store M samples into the delay line for each output you calculate, calculate the decimated output as the sum of products of the samples values and the filter coefficients, put next M samples in the pipeline, and repeat the decimation.

High-Pass Filters

The high-pass filter frequency response is as shown in Figure 8.1b. Now the only nonzero values are from $-\pi$ to $-\Omega_1$ and from Ω_1 to π. Integrating Equation 8.4 for the outer limit $-\pi$ to π and subtracting the zero part from $-\Omega_1$ to Ω_1 would result in the high-pass filter response. This is as if the low-pass filter is subtracted from 1. See Equation 8.10.

$$H_{HP}(\Omega) = 1 - H_{LP}(\Omega) \qquad (8.10)$$

The corresponding impulse response is,

$$h_{hp}(n) = \delta(n) - h_{lp}(n)$$

Where $\delta(n) = 1$ when $n = 0$, and 0 otherwise.

Substituting the value of $h(n)$ from the derivative of Equation 8.10 we get high-pass filter coefficients,

$$h_{hp}(n) = 1 - \frac{\Omega_1}{\pi} \times \frac{\sin(n\Omega_1)}{n\Omega_1} \quad \text{for } n = 0$$

$$h_{hp}(n) = -\frac{\Omega_1}{\pi} \times \frac{\sin(n\Omega_1)}{n\Omega_1} \quad \text{for } n \neq 0$$

The high-pass coefficients could also be obtained by subtracting the coefficients of the low-pass filter from the Nyquist frequency coefficients.

$$(\omega_c)_{HP} = (\omega_s/2) - (\omega_c)_{LP}$$

$$h_{hp}(n) = (-1)^n h(n)_{LP}$$

The following is an example of high-pass filter design.

EXAMPLE 8.2

Design a high-pass filter with the cutoff frequency of 3/4ths of the Nyquist frequency.

Computed values of 21 coefficients for a high cutoff frequency of 0.6π ($\Omega_1 = 0.3\pi$) are shown in Table 8.2 for $n = -10$ to $n = +10$.

The following are Matlab instructions for generating the filter coefficients:

```
p = ones (size(1:21)); p(2:2:20) = -1
c = p.*(0.25* sinc(0.25*(-10:10)));
```

TABLE 8.2 The Filter Coefficients for the High-pass Filter with the Cutoff Frequency of 0.6π.

h(−10)=h(10)	0.031831
h(−9)=h(9)	−0.025009
h(−8)=h(8)	0
h(−7)=h(7)	0.032154
h(−6)=h(6)	−0.053052
h(−5)=h(5)	0.045016
h(−4)=h(4)	0
h(−3)=h(3)	−0.075026
h(−2)=h(2)	.15915
h(−1)=h(1)	−0.22508
h(0)	.25

The convolution equation of the high-pass filter coefficients with the input samples is presented in Equation 8.11.

$$y[k] = \sum_{n=0}^{20} h_{HP} x[k-n] \qquad (8.11)$$

The frequency response is as shown in Figure 8.6.

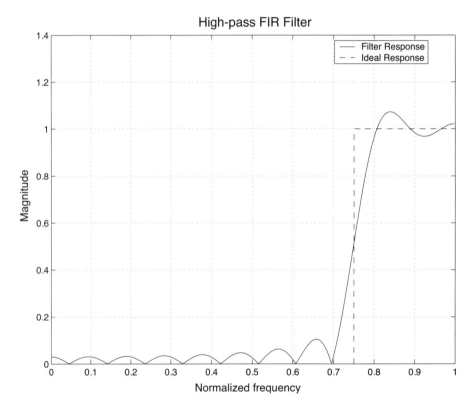

FIGURE 8.6 The magnitude response of the 21-tap high-pass filter (cutoff frequency $+\Omega = \frac{3\pi}{4}$) compared to the ideal filter.

Band-Pass Filters

The ideal band-pass filter frequency response is shown in Figure 8.1c. The band-pass center frequency is Ω_0 and $\Omega_1 = \Omega_0 - \Omega_c$, $\Omega_1 = \Omega_0 - \Omega_c$. The cutoff frequency Ω_c refers to the low-pass filter of Figure 8.1a. The corresponding impulse response is obtained from the appropriate integration of Equation 8.4, as shown in Equation 8.12.

$$h_{BP}(\pi) = (2\cos n\Omega_0)h(n)_{LP} \quad \text{for } n - 0, \pm1, \pm2, \ldots \tag{8.12}$$

The following example demonstrates the design of a band-pass filter using filter coefficients of Equation 8.12.

EXAMPLE 8.3

Design a band-pass filter with the low cutoff frequency 1/4th of the sampling frequency and the high cutoff frequency 3/4ths of the sampling frequency.

The band-pass filter of ($\Omega_1 = 0.25\pi$) and the high cutoff frequency of ($\Omega_2 = 0.75\pi$) are shown in Table 8.3 for $n = -10$ to $n = +10$. (See Appendix B for the Scilab listing to generate the coefficients.)

The following are Matlab instructions for generating the filter coefficients:

```
>>b = 0.25*sinc(0.25*[-10:10]);
>>c = 2*cos([-10:10]* pi/2)/*b
```

TABLE 8.3 The Filter Coefficients for a Band-Pass Filter for $n = -10$ to $n = +10$

h(−10)=h(10)	−0.063662
h(−9)=h(9)	0
h(−8)=h(8)	0
h(−7)=h(7)	0
h(−6)=h(6)	.1061
h(−5)=h(5)	0
h(−4)=h(4)	0
h(−3)=h(3)	0
h(−2)=h(2)	−.31831
h(−1)=h(1)	0
h(0)	−.5

And the filter output is shown in Equation 8.13.

$$y[k] = \sum_{n=0}^{20} h_{BP} x[k-n] \quad (8.13)$$

The frequency response of Equation 8.13 is as shown in Figure 8.7.

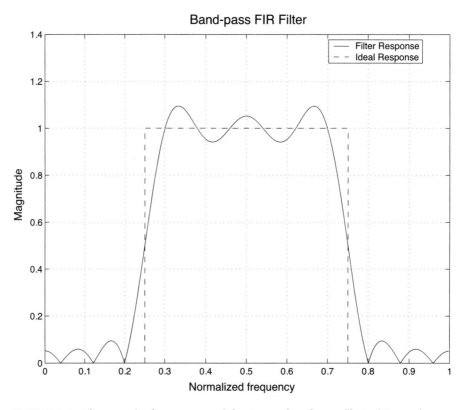

FIGURE 8.7 The magnitude response of the 21-tap band-pass filter of Example 8.3.

Band-Stop Filters

We can obtain an expression for a band-stop filter from the coefficients of band-pass filters by simply negating the coefficients everywhere except at 0, just like the high-pass filter obtained from the low-pass filter. The coefficient at 0 is a high-band pass coefficient subtracted from 1. See Equation 8.14.

$$h_{BS}(n) = 1 - h_{BP} \quad \text{for } n = 0$$
$$h_{BS}(n) = -h_{BP} \quad \text{for } n \neq 0 \quad (8.14)$$

EXAMPLE 8.4

Design a band-stop filter with the low cutoff frequency 1/4th of the sampling frequency and the high cutoff frequency 3/4ths of the sampling frequency.

The band-pass filter of ($\Omega_1 = 0.25\pi$) and the high cutoff frequency of ($\Omega_2 = 0.75\pi$) are shown in Table 8.4 for $n = -10$ to $n = +10$. Please see Appendix B for the Scilab listing to generate the coefficients.

The following are Matlab instructions for generating the filter coefficients:

```
>>b = 0.25*sinc(0.25*[-10:10]);
>>t = 2*cos([-10:10]*pi/2).*b
>>h0 = 1 - (t(11:11));
>>t = -t;
>>t(11:11) = h0;
```

TABLE 8.4 The Filter Coefficients for a Band-Pass Filter for $n = -10$ to $n = +10$

h(−10)=h(10)	0.063662
h(−9)=h(9)	0
h(−8)=h(8)	0
h(−7)=h(7)	0
h(−6)=h(6)	−.1061
h(−5)=h(5)	0
h(−4)=h(4)	0
h(−3)=h(3)	0
h(−2)=h(2)	.31831
h(−1)=h(1)	0
h(0)	−.5

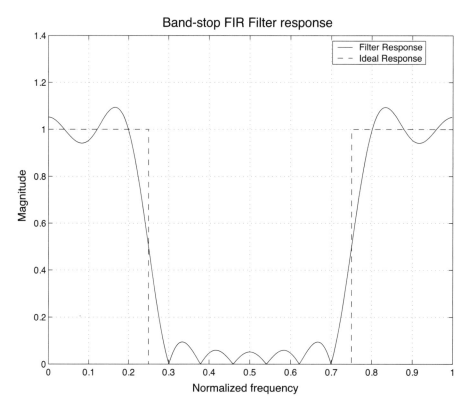

FIGURE 8.8 The magnitude response of the 21-tap band-stop filter (low cutoff frequency $+\Omega = \frac{\pi}{4}$), (high cutoff frequency $+\Omega = \frac{3\pi}{4}$), compared to the ideal filter.

A plot of the filter Transfer Function of Equation 8.14 is shown in the Figure 8.8.

The FIR filters derived from the ideal response always produce ripples in the pass-band region. These ripples are the result of truncating the infinite coefficients Fourier series while ignoring sinusoids in the process of summation. Increasing the number of filter coefficients improves the ripples somewhat (see Figure 8.9 for a comparison of 11, 21, and 31 coefficients Transfer Function). These ripples are unavoidable, but there are ways to improve the design and that is the topic of discussion of our next section.

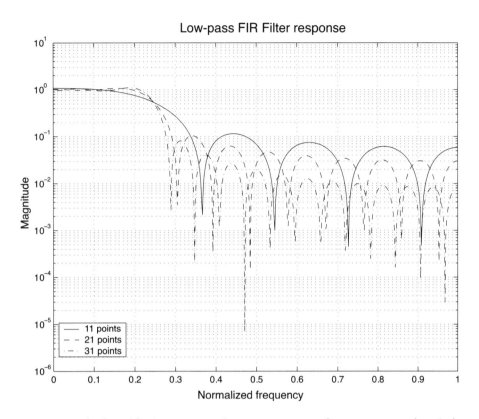

FIGURE 8.9 The logarithmic response of 11-tap, 21-tap, and 31-tap rectangular windows.

WINDOWING

Frequency domain multiplication is akin to time domain convolution; similarly, time domain multiplication is the same as frequency domain convolution. Performing the Fourier Transform on a set of data is like having a window to a different domain. From the window of convolution, we see frequencies and through frequency domain multiplication, we see time series. We see one through the other. Imagine (when performing the infinite summation) you are looking through a convolving window, and you see all the frequencies that are present in the impulse response. But a truncated series would miss some frequencies and would appear as ripples in the Transfer Function. It has the same effect as multiplying an infinite impulse response with a rectangular window of 1s and 0s. In our example of filters of 21 coefficients, we were forced to truncate the infinite number of coefficients into a fixed number of 21 values. We were by default using a rectangular window of 21 1s and the rest 0s as shown in Figure 8.10.

$$0,0,0,...0,0,0,1,1,1...1,1,1,0,0,0..0,0,0$$

The Fourier series equivalent of the ideal frequency response of Figure 8.1b is given as,

$$g(\Omega) = \sum_{k=-\infty}^{k=\infty} c_k e^{jk\Omega}$$

The coefficients c_k represent the amplitudes of the impulse response. But for practical consideration, the infinite summation must be truncated to a finite value,

$$c_k = \frac{1}{2\pi} \int_0^n f(t) e^{-jk\Omega} dt$$

The affect of multiplying the c_k values with a series of $(2N + 1)$ nonzero terms can be described graphically, as shown in Figure 8.11. The number of sample points $(2N + 1)$ merely indicates a symmetric window with N number of sample points on each side of the center.

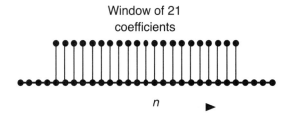

FIGURE 8.10 Discrete time rectangular window of 21 points.

The window determines how much of the impulse response can be seen. Looking from the multiplication window, an infinite length impulse response $h(n)$ (essentially, an inverse Fourier Transform of an ideal frequency response) when multiplied with a rectangular window $w(n)$ produces a truncated impulse response $t(n)$. This is a time domain multiplication, but the effect is the same as frequency domain convolution, and that is what we obtained when we realized the Transfer

Function through a truncated series of impulse response; see Figure 8.11. The result of chopping of the ends of the infinite series shows up as ripples in the Transfer Function near the transition band as well as side lobes in the stop band.

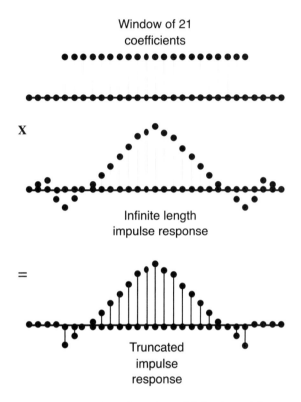

FIGURE 8.11 Time domain multiplication of the rectangular window with the infinite impulse response of the ideal infinite rectangular window.

Ideally, if the rectangular convolving window were as large as the frequency response window, there wouldn't be any distortion of the resulting Transfer Function. Looking from another angle, the process of obtaining the actual impulse response $H_A(\Omega)$ is as if we have done a frequency domain convolution of a rectangular convolving window $W(\Omega)$ with an ideal frequency response $H_I(\Omega)$ as shown in Equation 8.15.

$$H_A(\Omega) = H_I(\Omega) * W(\Omega) \tag{8.15}$$

Help from Matlab and Scilab

Matlab and Scilab have built-in functions that compute the filter coefficients, and you can specify the window type to be used as a parameter to the function. Appendix A, "Matlab Tutorial," provides a tutorial on creating an FIR filter using Matlab functions, and it is recommended that you exercise the examples given in the tutorial.

The Rectangular Window

We characterize a rectangular window as a series of $(2N + 1)$ nonzero terms. The values are either 1 (inside the window) or 0 (outside the window). It is convenient to have a window symmetric on 0, thus we have N sample points on each side. We can always shift the window so that it begins with 0 to make it a causal window.

$$0,0,0,..0,0,0,1,1,1...1,1,1,0,0,0..0,0,0$$

For $2N + 1$ number of 1s, the window is as shown in Equation 8.16.

$$a_k = \begin{cases} 1 & |k| \leq |N| \\ 0 & |k| > |N| \end{cases} \quad (8.16)$$

The Fourier Transform of the rectangular window of Equation 8.16 is shown in Equation 8.17.

$$W_R(\Omega) = \sum_{k=-N}^{k=N} e^{jk\Omega} = e^{-jN\Omega} + e^{-j(N-1)\Omega} + \cdots + e^{j(N-1)\Omega} + e^{jN\Omega} \quad (8.17)$$

Equation 8.17 is a polynomial evaluated on a unit circle whose coefficients are 1, and Figure 8.5 is its frequency response plot where $N = 10$. The Scilab instructions for evaluating the Transfer Function magnitude of Equation 8.17 are provided in Appendix B.

Figure 8.12 is the magnitude of Equation 8.17 plotted for different values of N. Notice the reduction in transition width as we increase the number of coefficients. But the side lobe amplitude remains same. The side lobe amplitude is specific to a window function and is independent of the number of coefficients in the window. For a rectangular window, the first side lobe amplitude is around 13 dB, which is about 1/5th of its pass-band magnitude. In most situations, it is probably not an acceptable value, but compared to other windows, the rectangular window offers a steeper transition width.

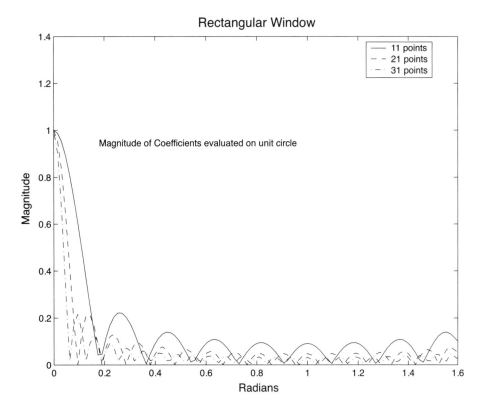

FIGURE 8.12 The plot showing the magnitude response of the rectangular window for a different number of filter coefficients.

There is an approximate relationship between the transition widths and the number of coefficients, and for a rectangular window, it is $(2N+1) \times \Delta f \approx 1.9$. Two aspects are important when specifying a filter,

- The transition width
- The side lobe amplitude

The number of coefficients affects the transition width, but the side lobe amplitude is fixed for a window type.

EXAMPLE 8.5

Suppose a low-pass filter is specified as

$$0.8 \leq |H(e^{j\omega})| \leq 1.2 \quad 0 \leq \omega \leq 0.18\pi$$

$$|H(e^{j\omega})| \leq 0.2 \quad 0.22 \leq \omega \leq \pi$$

The requirements of a magnitude drop of 0.2 is within the reach of the rectangular window, and we can compute the number of coefficients for the transition width of 0.04 as,

$$(2N+1) \times 0.04 \approx 1.9$$

$$N \approx 23$$

The magnitude summation of Equation 8.17 can also be expressed as shown in Equation 8.18.

$$W_R(\Omega) = \frac{e^{j(N+1/2)\Omega} - e^{-j(N+1/2)\Omega}}{e^{j\Omega/2} - e^{-j\Omega/2}}$$

$$W_R(\Omega) = \frac{\sin(N+1/2)\Omega}{\sin(\Omega/2)} \tag{8.18}$$

If we were to *normalize* the window function to a unit area (meaning the sum of the coefficients to be equal to 1), then each coefficient is an equal weight of $\frac{1}{2N+1}$ and Equation 8.18 becomes Equation 8.19:

$$a_k = \begin{cases} \dfrac{1}{2n+1} & |k \leq N| \\ 0 & |k > N| \end{cases} \tag{8.19}$$

And the corresponding impulse response of Equation 8.19 becomes,

$$W_R(\Omega) = \frac{\sin(N+1/2)\Omega}{(1/2N+1)\sin(\Omega/2)}$$

All the filters discussed in the previous section were designed using the rectangular window, as the coefficients were generated with the values from the ideal rectangular transfer function. Next, we discuss other window functions that have better suppression in the stop-band region compared to the rectangular window.

The Power Window

Shaping the rectangular window with a power function multiplier can make a simple design improvement. In fact, any function whose values tapers off toward the end can serve the purpose of a window shaper and provide a performance improvement. A function that puts increasingly extra weight towards the end is shown in Equation 8.20.

$$a_k = \begin{cases} (1-(k/N)^2) & |k| \le |N| \\ 0 & |k| > |N| \end{cases} \qquad (8.20)$$

The maximum weight of the power function of Equation 8.20 is 1 when $k = 0$ and for $k = N$ the power function tapers off to 0 values. See Equation 8.21. The coefficients are listed in Table 8.5.

$W_{PW}(0) = 1 \qquad\qquad W_{PW}(0) = 1$
$W_{PW}(1) = .99 \qquad\qquad W_{PW}(11) = .99$
$W_{PW}(2) = .96 \qquad\qquad W_{PW}(12) = .96$
$W_{PW}(3) = .91 \qquad\qquad W_{PW}(13) = .91$
$W_{PW}(4) = .84 \qquad\qquad W_{PW}(14) = .84$
$W_{PW}(5) = .75 \qquad\qquad W_{PW}(15) = .75$
$W_{PW}(6) = .64 \qquad\qquad W_{PW}(16) = .64$
$W_{PW}(7) = .51 \qquad\qquad W_{PW}(17) = .51$
$W_{PW}(8) = .36 \qquad\qquad W_{PW}(18) = .36$
$W_{PW}(9) = .19 \qquad\qquad W_{PW}(19) = .19$
$W_{PW}(10) = 0 \qquad\qquad W_{PW}(20) = 0$

$$h(n) = \frac{1}{2\pi} \int_{-\Omega_1}^{\Omega_1} (1-(n/N)^2) \bullet 1 \bullet e^{jn\Omega} d\Omega = \frac{(1-(n/N)^2) e^{jn\Omega}}{2\pi \; jn} \Bigg]_{-\Omega_1}^{\Omega_2}$$

$$h(n) = \frac{(1-(n/N)^2)}{n\pi} \left(\frac{e^{jn\Omega} - e^{-jn\Omega}}{2j} \right) = \frac{(1-(n/N)^2)\sin(n\Omega_1)}{n\pi}$$

$$h(n) = \frac{\Omega_1(1-(n/N)^2)}{\pi} \times \frac{\sin(n\Omega_1)}{n\Omega_1} \qquad (8.21)$$

TABLE 8.5 Coefficients for a 21-Tap Low-Pass Filter (Cutoff Frequency $\Omega_c = 0.4$)

h(−10)=h(20)	0
h(−9)=h(19)	−.0063910
h(−8)=h(18)	−.0084194
h(−7)=h(17)	.0136314
h(−6)=h(16)	.0322913
h(−5)=h(15)	0
h(−4)=h(14)	−.0635734
h(−3)=h(13)	−.0567530
h(−2)=h(12)	.0898070
h(−1)=h(11)	.2997034
h(0)	.4

Figure 8.13 is the magnitude of Equation 8.21 plotted for different values of N, and the Scilab instructions for evaluating the Transfer Function magnitude are provided in Appendix B. There is some improvement in the side-lobe amplitude as it dropped down to nearly 21 dB but the transition width goes up from 1.9 to 2.8 for the same number of coefficients. It only indicates that to meet the requirement of steeper roll-off of frequency response, we must increase the number of coefficients.

The power window coefficients may be created with the help of Matlab. The following instructions create 21 coefficients as shown in Table 8.5 for a filter with the cutoff frequency $\Omega_c = 0.4$.

```
>> N=-10:10
>> pc = (1-(N./10).^2)
>> b =pc.*( 0.4*sinc(0.4*(N)))
```

The coefficients in the variable b are the coefficients of the numerator polynomial of the filter transfer function. Since there is no feedback in the FIR filters, we could leave the denominator polynomial simply as 1. The following instructions apply the power window filter to a random set of data.

```
>> w=randn(1000,1);
>> y=filter(b,1,w);
```

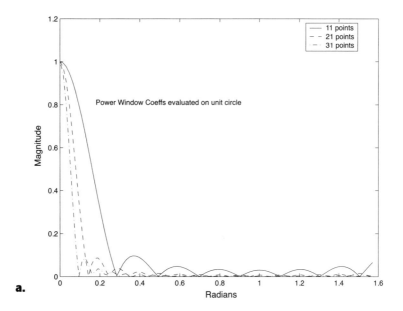

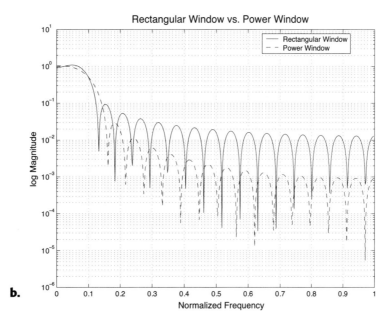

FIGURE 8.13 a) The plot showing the magnitude response of the power window for different filter coefficients. b) Reduction in the side lobe amplitude with the power window compared to the rectangular window.

Figure 8.14 displays the power spectrum of the data, before and after the filtering. The top figures are the random sequence and its power spectrum showing distribution of frequencies in all bands. The bottom figures are the filtered output and the power spectrum. You could see a drop in the frequencies beyond 200 Hz.

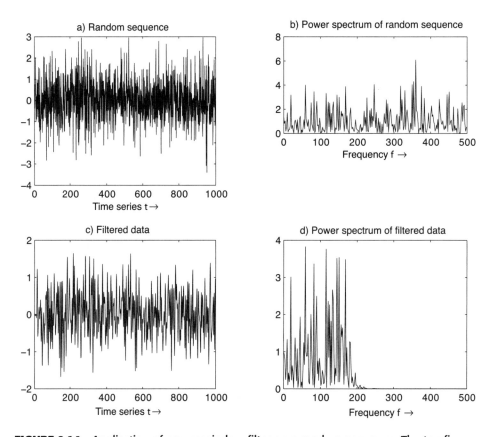

FIGURE 8.14 Application of power window filter on a random sequence. The top figures are the random sequence and its power spectrum; the bottom figures are the filtered output and its power spectrum. Notice the reduction in higher frequency component.

The following example illustrates the difference between rectangular window and the power window for the number of coefficients required for the same transition width as in the rectangular window.

EXAMPLE 8.6

Suppose a low-pass filter is specified as

$$0.1 \le |H(e^{j\omega})| \le 1.1 \qquad 0 \le \omega \le 0.18\pi$$

$$|H(e^{j\omega})| \le 0.1 \qquad 0.22 \le \omega \le \pi$$

The requirements of a magnitude drop of 0.1 is within the reach of the power window, and to achieve the transition width of 0.04, we get,

$$(2N+1) \times 0.04 \approx 2.8$$

$$N \approx 35$$

Figure 8.15 is the plot of the filter response of Example 8.1 using Hamming window coefficients (together with rectangular window for comparison). There is a 3% overshoot near the transition band, but the filter has a steeper response compared to other window functions, see Figure 8.21.

The Von Hann Window

A more severe weighting average that would put extra weight at the end is a cosine function, given in Equation 8.22. This is the Von Hann window or the raised cosine window,

$$a_k = \begin{cases} \dfrac{1+\cos\pi k/N}{2} & |k \le N| \\ 0 & |k > N| \end{cases} \qquad (8.22)$$

The following are the Matlab instructions for creating 21 coefficients in column vectors for a filter with cutoff frequency $\Omega_c = 0.4$.

```
>> b =( 0.4*sinc(0.4*(N))
>> b = b.*hann(21)';
```

The results are tabulated for $k = 0$ through 10 only; being a symmetric window, the values for $k = -1$ to -10 are same as $k = 1$ through 9.

$$W_{PW}(0) = 1$$
$$W_{PW}(1) = .977486$$
$$W_{PW}(2) = .91214782$$
$$W_{PW}(3) = .81038122$$
$$W_{PW}(4) = .678$$
$$W_{PW}(5) = .54$$
$$W_{PW}(6) = .3978$$
$$W_{PW}(7) = .2696$$
$$W_{PW}(8) = .172$$
$$W_{PW}(9) = .1025$$
$$W_{PW}(10) = 0.08$$

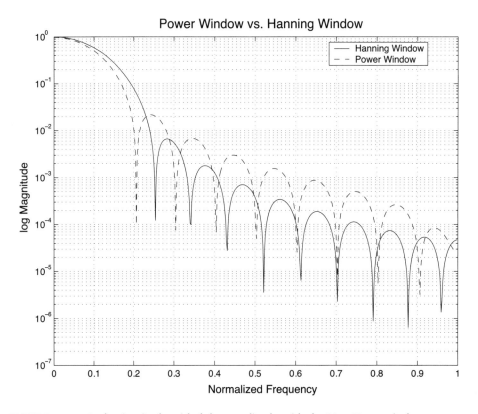

FIGURE 8.15 Reduction in the side lobe amplitude with the Von Hann window compared to the power window.

The magnitude of Equation 8.22 plotted for different values of *N* is presented in Figure 8.16. (The Scilab instructions are provided in Appendix B.) The side lobe amplitude drops down to nearly 31 dB but compared to the rectangular window, the transition width is up from 1.9 to 4.2.

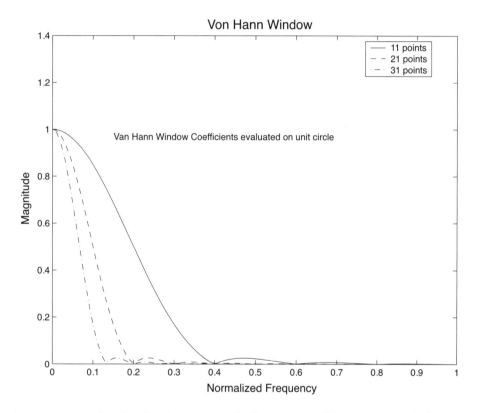

FIGURE 8.16 The plot showing the magnitude response of the Von Hann window for a different number of filter coefficients.

The filter specifications of Example 8.6 are met with 52 coefficients for the Von Hann window compared to 23 coefficients of the rectangular window, but on the bright side, we do get a suppression of 31 dB in the stop-band region.

$$(2N+1) \times 0.04 \approx 4.2$$

$$N \approx 52$$

The Hamming Window

The Hamming Window, also known as cosine window with platform, is given in Equation 8.23.

$$a_k = \begin{cases} 0.54 + (1-0.54)\cos \pi k/N & |k \le N| \\ 0 & |k > N| \end{cases} \quad (8.23)$$

The following are the Matlab instructions for creating 21 coefficients in column vectors for a filter with cutoff frequency $\Omega_c = 0.4$.

```
>> b =( 0.4*sinc(0.4*(N))
>> b = b.*hamming(21)';
```

Apply the transfer function to input sequence x,

```
>> y = filter(b,1,x);
```

The following table provides the values of Equation 8.23 for $k = 0$ through 9,

$$W_{PW}(0) = 1$$
$$W_{PW}(1) = .9755$$
$$W_{PW}(2) = .9045$$
$$W_{PW}(3) = .7938$$
$$W_{PW}(4) = .6545$$
$$W_{PW}(5) = .5$$
$$W_{PW}(6) = .3454$$
$$W_{PW}(7) = .206$$
$$W_{PW}(8) = .095$$
$$W_{PW}(9) = .0244$$
$$W_{PW}(10) = 0$$

Figure 8.17 shows the magnitude of Equation 8.23 plotted for different values of N with the help of the Scilab program (see Appendix B). The side lobe amplitude drops down to nearly 41 dB, but compared to rectangular window, the transition width goes up from 1.9 to 4.4.

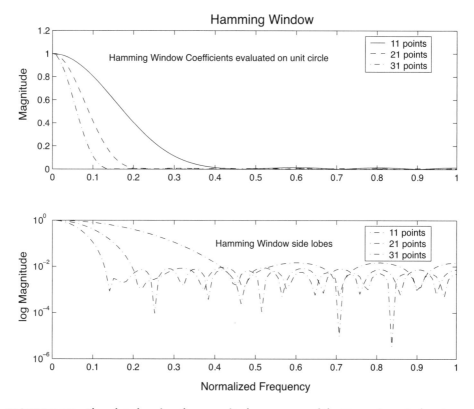

FIGURE 8.17 The plot showing the magnitude response of the Hamming window for a different number of filter coefficients.

Compared to the rectangular window, a transition width of 0.4 requires

$$(2N + 1) \times 0.04 \approx 1.8$$

$$N \approx 55$$

A plot of the filter response of Example 8.1 modified by the Hamming window is presented in Figure 8.18 (together with rectangular window for comparison). There is a 0.03% overshoot near the transition band and the side lobe suppression is more aggressive, but there is an increase in the width of the roll-off of the transition band. (See Figure 8.21 for comparison.)

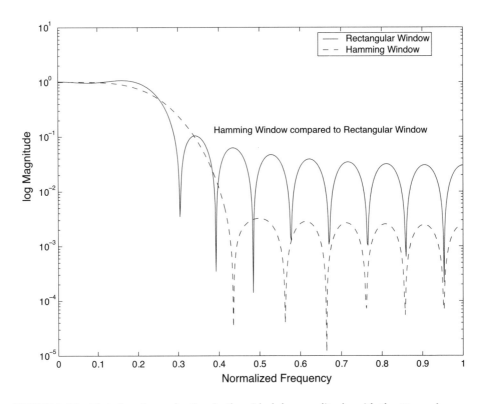

FIGURE 8.18 Plot showing reduction in the side lobe amplitude with the Hamming window compared to the rectangular window.

Kaiser Window

The Kaiser window's weighting function has a flexible design option. You can specify a range of allowable ripples in the side lobes (those were fixed in the other window options). The function is given in Equation 8.24.

$$a_k = \begin{cases} \dfrac{I_0\left[\beta\sqrt{1-(n/N)^2}\right]}{I_0(\beta)} & |k \leq N| \\ 0 & |k > N| \end{cases} \quad (8.24)$$

$$I_0 = 1 + \sum_{m=1}^{M}\left(\dfrac{(x/2)^m}{m!}\right)^2$$

The parameter I_0 is the modified Bessel function of the first kind and zero order. Equation 8.24 approximates the Bessel Function where summation limit M is an arbitrary limit. We can stop computing the Bessel terms when the value becomes less then a prescribed low limit. The Scilab program for computing Kaiser Window coefficients using Equation 8.24 is presented in Appendix B (see function Kaiser_coeff()).

The β value specifies its frequency response in terms of the main lobe width and the side lobe level. Large value of β corresponds to a wider main lobe with smaller side lobe levels. The normal range of β is $4 \leq \beta \leq 9$. The parameter M is chosen such that the last term in series is less than 1×10^{-8}. The following table was computed using the value $\beta = 2\pi$. (See Appendix B.)

$$W_{KR}(0) = 1$$
$$W_{KR}(1) = .9801$$
$$W_{KR}(2) = .9216$$
$$W_{KR}(3) = .8281$$
$$W_{KR}(4) = .7056$$
$$W_{KR}(5) = .5625$$
$$W_{KR}(6) = .4096$$
$$W_{KR}(7) = .2601$$
$$W_{KR}(8) = .1296$$
$$W_{KR}(9) = .0361$$
$$W_{KR}(10) = 0$$

R. W. Hamming has summarized the empirical relationship between the parameter β and the stop-band ripples α in dB as follows,

$$\beta = \begin{vmatrix} 0.1102(\alpha - 8.7) & \alpha > 50 \\ 0.5842(\alpha - 21)^{0.4} \_ 0.07886(\alpha - 21) & 21 \leq \alpha \leq 50 \\ 0 & \alpha < 21 \end{vmatrix}$$

And the relationship between the transition width and the number of coefficients for is given as,

$$(2N+1) = \frac{\alpha - 7.95}{14.36 \Delta f}$$

For a stop-band ripple of 0.01 and transition width of 0.01*f*, we could calculate the number of coefficients as,

$$\alpha = 20\log(0.01) = -40$$

$$(2N+1) = \frac{40 - 7.95}{14.36 \times 0.01} = 223$$

Figure 8.19 is the magnitude of Equation 8.24 plotted for different values of N. The relationship between the number of coefficients and the transition width for $\beta = 2\pi$ is given as $(2N + 1) \times \Delta f \approx 0.44$.

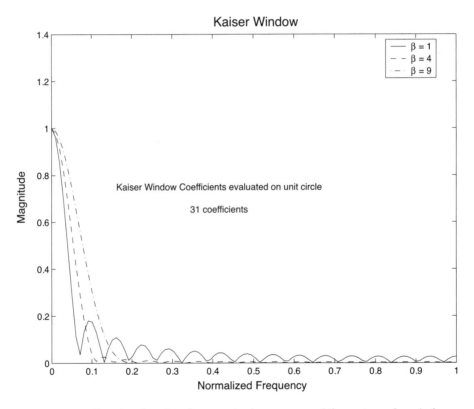

FIGURE 8.19 The plot showing the magnitude response of the rectangular window for a different number of filter coefficients.

Figure 8.20 is a plot of the filter response of the Kaiser window function of Equation 8.24 compared to the rectangular window.

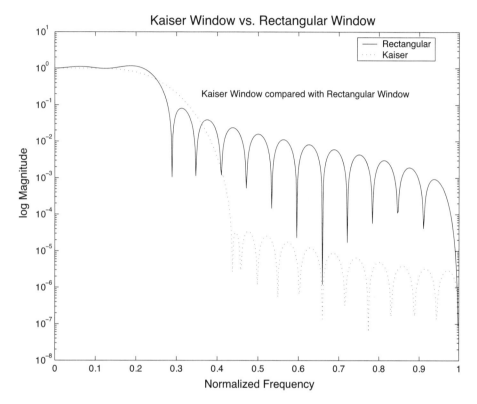

FIGURE 8.20 Reduction in the side lobe amplitude with the Kaiser window compared to the rectangular window.

A comparison of window functions shows that applying window function is a better alternate than increasing the coefficients of a rectangular window. Figure 8.21 is a Bode plot of the response of the four window functions as mentioned in the previous section.

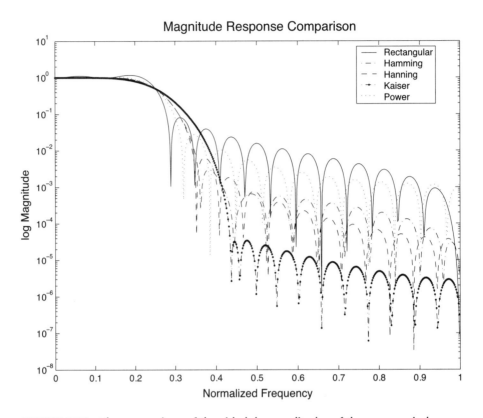

FIGURE 8.21 The comparison of the side lobe amplitudes of the power window, Von Hann window, Hamming window and the Kaiser window.

SUMMARY

We have studied the Fourier Transform method of designing filters, commonly known as Finite Impulse Response (FIR) filters. The method provides filters that do not depend upon values from the past, and the impulse response has a finite duration, thus, providing a linear phase response compared to the Infinite Impulse Response method in which the initial transients stay on for a long time.

Appendix A: Matlab Tutorial

Matlab is a powerful interactive programming environment with an enormous list of built-in functions for numerical computations and data analysis. It is a comprehensive tool in the hands of engineers and scientists who are interested in signal processing and digital filtering functions.

It is an interpreter and a high-level programming language offering:

- 2D and 3D graphics, animation
- Linear algebra, sparse matrices
- Polynomials and rational functions
- Simulation: Ordinary Differential Equation solver
- Dynamic systems modeler and simulator
- Classic and robust control, LMI optimization
- Differentiable and nondifferentiable optimization
- Signal processing
- Statistics

Matlab is basically a library of programs designed to operate upon matrices. The goal of this tutorial is to provide a hands-on introduction to Matlab. It is assumed that the Matlab with Signal Processing and Filter Design Toolbox software is already installed and you can enter commands through its command-line interpreter as well as execute scripts by entering the scripts' names.

The graphs and figures in this book were prepared using Matlab scripts, and the listings are provided in the directory Matlab on the accompanying CD-ROM. You may use these scripts as examples for practice.

Help is always available on the shell window, and a search engine is provided for a search on any topic. Get used to Matlab help, as it is extensive and offers lots of examples to try.

307

EXECUTING SCRIPTS

You can enter the commands either through the keyboard or execute a Matlab script by simply entering the name of the script on the command line, without the .m suffix. The Matlab interpreter displays a prompt for you to enter the commands,

```
>>
```

For example, to display Figure 8.14, simply enter the following command at the Matlab prompt,

```
>> cd cdrom/matlab
>> f8_14
```

Importing Data from a File

You may use the `load` command to import data into existing variable space. By default, Matlab creates a variable with the same name as the filename, minus the file extension.

```
>> load mdata.txt;
```

If you want to assign data a variable name other than the filename, use the functional form.

```
>> Z=load('mdata.txt');
```

If your data contains delimiters other than space, you may use the `dlmread` command that takes an argument specifying the delimiter.

```
>> Z = dlmread ('mdata.txt',';')
```

Exporting Data to a File

You may use the `save` command to export data into a file, specifying -ASCII to save the file in ASCII format.

```
>> save mdata.txt A -ASCII
```

A better option of save is `dlmwrite`, which takes a delimiter argument.

```
>> dlmwrite ('mdata.txt',Z,';')
```

Use the `diary` command to export all the commands executed in a Matlab session until it is turned off.

```
>> diary m.out
.
.
>> diary off
```

Operations on Matrices

When you enter commands through the keyboard, Matlab displays the values on the screen. The values of the rows in a matrix are entered with spaces between them, and values of the columns of a matrix are entered with semicolons between them. For example, to assign a 3 × 2 matrix A, with two rows and three columns, enter the data as shown followed by the carriage return,

```
>> A = [4 5 6; 1 2 3]
```

The result will be displayed as,

```
>>
A =
     4    3    2
     1    2    3
>>
```

You can suppress the result display by terminating command with a semicolon,

```
>> A = [4 3 2; 1 2 3];
>>
```

Obtain a 3-by-3 matrix with symmetric entries,

```
>> A = pascal(3)
```

```
A =
    1    1    1
    1    2    3
    1    3    6
```

A matrix of magic square:

```
>> B = magic(3)
B=
    8    1    6
    3    5    7
    4    9    2
```

The Transpose operation interchanges rows to columns

```
>> X = B'
X=
    8    3    4
    1    5    6
    6    7    2
```

A 3-by-2 rectangular matrix of random integers:

```
>> C = fix(10*rand(3,2))

C =
    2
         3    5
    5
```

The row and column of a matrix are treated as vectors. To obtain a *column sum* of the matrix columns enter,

```
>> sum(A)

    5    5    5
```

To obtain a *row sum*, you may have to transpose the matrix, as Matlab by default works on matrix columns

```
>> A=A'

1
2
3
>> sum(A)

    9    6
```

A matrix with all ones in the diagonal and zeros elsewhere is an identity matrix. Use eye to obtain an identity matrix of the size *n*-by-*m*.

```
>> P=eye(3,2)

P =
0
1
0
```

To obtain the diagonal element of a matrix:

```
>> diag(A)
```

To flip a matrix from left to right,

```
>> fliplr(A)
```

To extract the element of a matrix, enter from row i and column j, enter A(i,j),

```
>> A(3,2)

ans =
    3
```

Get matrix of all zeros

```
>> Z = zeros(2,3)
ans =
    0    0    0
    0    0    0
```

Get matrix of all ones

```
>> Z = ones(2,3)
ans =
     1     1     1
     1     1     1
```

Get matrix of uniformly distributed random elements and round towards zero

```
>> Z = fix(10*rand(2,3))
ans =
     9     1     9
     7     4     9
```

Get matrix of normally distributed random elements and round towards zero

```
>> Z = randn(2,3)
ans =

   -0.0956    0.2944    0.7143
   -1.3362    0.8323    1.1012
```

Join small matrices to make bigger ones

```
>> A = [1 2 3];
>> B = [4,5,6];
>> C = [A B]

ans =
     [1    2    3    4    5    6]
```

Multiply all elements of a matrix with a scalar

```
>> A=[1 2; 2 3];
>> 8.*A

ans =
16
24
```

Divide all elements of a matrix with a scalar

```
>> A=[1 2; 2 3];
>> A./8

ans =

0.2500       0.2500
0.2500       0.3750
```

Delete column 1 of all rows

```
>> A(:,1) = []
```

Delete row 1 from the matrix

```
>> A(1,:) = []
```

Two matrices with dimension m-by-p and p-by-n may be multiplied to produce an m-by-n matrix. The * is the matrix multiply operator.

```
>> X = A*B;
```

Complex Numbers

The real part of the complex number is entered as regular numbers, but the imaginary part (with the Gaussian operator $\sqrt{-1}$) may be entered with the suffix i or j.
   Matrix with complex values

```
>> A = [1+4i, 2-3i, 4+3j]

A=
     1+ 4.000i    2 - 3.000i    4 + 3i
```

Extract only the real values of a matrix

```
>> real(A)
ans=
     1    2    4
```

Extract only the imag values of a matrix

```
>> imag(A)
ans=
         4        -3         3
```

Get absolute values of complex numbers in a matrix

```
>> abs(A)
ans=
         3.1623            4.4721            6.4031
```

Determinant of a square matrix

```
>> B = [1 , 2+3i; 4+3i, 2];
>> det(B)

ans=
-18.0000i
```

Inverse of a matrix

```
>> inv(B);
```

Eigenvalues of a matrix

```
>> eig(B);
```

Build a table of mean and variance; the following example creates a matrix of 5 rows and 2 columns and then computes the mean and variance

```
>> D=[rand(5,1), rand(5,1)];
>> mean(D)

ans =
0.4615
```

Complex conjugate

```
>> z_bar = conj(1+2i);
```

THE COLON OPERATOR

The : operator is used in several different ways:
Use : as an integer increment,

```
>> 1:8
ans =
     1     2     3     4     5     6     7     8
```

Use : as nonunit space between a starting and end value

```
>> 1:.5:3
ans =
    1.0000  1.5000  2.0000  2.5000  3.0000
```

Use : to refer a portion of a matrix,

```
>> A(1:2,1)
ans =
     4
     3
```

Use : to refer to all row elements of a matrix, for example, to get all row values in the first column, enter

```
>> A(:,1)


ans =
     4
     1
```

Use : to refer to all column elements of a matrix, for example, to get all column values in the first row, enter

```
>> A(1,:)
ans =
     4     3     2
```

EXPRESSION AND SPECIAL CONSTANTS

Matlab accepts regular expression operators such as those listed in Table A.1.

TABLE A.1 Regular Expression Operators Accepted by Matlab

+	Add
-	Subtract
*	Multiply
/	Divide
\	Left matrix divide (see polynomial least square fit)
^	Power
'	Complex conjugate transpose
( )	Order of evaluation
pi	3.14159
i	$\sqrt{-1}$
j	$\sqrt{-1}$
eps	Floating-point relative precision
realmin	Smallest floating-point number
realmax	Largest floating-point number
Inf	Infinity
NaN	Not a number
sqrt	Square root
abs	Absolute value
exp	e
log	Logarithm based e
log10	Logarithm based 10

OTHER DATA STRUCTURES

Other data structures are explained in the following sections.

Multidimensional Arrays

Arrays with more than two subscripts may be created by calling zeros, ones, rand, or randn statements.

```
>> R =randn(2,3,4);
```

Cell Arrays

Cell arrays are multidimensional arrays whose elements are copies of other arrays. Matlab's ans command displays the values only if there is room to print.

```
>> C = {R sum(R)}

ans =
        [2x3x4 double] [1x1x2 double]
```

Characters and Text

Use single quotes to assign text strings to a variable.

```
>> s = ['Hello',' World']

s =
        Hello World
```

The function double converts a character matrix to a numeric matrix containing a floating-point representation of the ASCII code,

```
>> a = double(s);
```

Conversion

The function char converts floating-point to ASCII code,

```
>>s = char(a);
```

The functions int8, int16, int32, and int64 convert floating-point numbers to integers,

```
>>s = int8(3.4);
ans =
        3
```

Structures

Structures are multidimensional arrays with field designators,

```
>> S.name = 'Test'

S =
name:   'Test'
 >>S.elements = 5;

S =
        Name:       Test
        Elements:   5
```

Functions

Functions are M-files that can accept input arguments and return output arguments. You can view the example functions provided in the directory toolbox/matlab/matfun with the following command:

```
>> type rank
```

eval

The `eval` function evaluates or executes the statements in the text string.

```
>> S = ['type rank'];
>> eval (S)
```

POLYNOMIALS

Polynomials are represented by an array of coefficients. For example, the polynomials $x^3 + 4x^2 + 9x + 2$ are entered as,

```
>> p = [1 4 9 2];
```

Matlab stores polynomial coefficients in row vectors and polynomial roots in column vectors. The roots of polynomials are obtained by

```
>> roots(p)
ans =
-4.0542
0.0271 + 1.4897i
- 1.4897I
```

From the roots, the coefficients are obtained as

```
>> p=poly([-4.0542; 0.0271+1.4897i; 0.0271-1.4897i;])
ans =
1    4     2.0002  9.0001
>>
```

Evaluating Polynomials

To evaluate the polynomial of the previous example over a domain of x that includes the roots, define the domain x

```
>> x = -10: .1: 10;
```

and evaluate the polynomial

```
>> y=polyval(p,x);
```

Multiplication of two polynomials is performed by convolution, for example,

$$(x^3 + 4x^2 + 9x + 2) \times (2x + 1) = 2x^4 + 9x^3 + 8x^2 + 20x + 9$$

```
>> p1 = [1 4 9 2];
>> p2 = [2 1];
>>p3=conv(p1,p2)
ans =
        2    9    8    20    9
```

The inverse operation polynomial division is performed by deconv, returning quotient and remainder.

```
>> [Q,R]= deconv(p3,p1)
ans
Q
          2    1
R
          0    0    0    0    0
```

SYSTEMS OF LINEAR EQUATIONS

Consider the following system of linear equations:

$$x_1 + 3x_2 + 7x_3 = 5$$
$$3x_1 + 2x_2 + 9x_3 = 13$$
$$4x_1 + x_2 + 9x_3 = 7$$

Defining the coefficient matrix A

$$A = \begin{bmatrix} 1 & 3 & 7 \\ 3 & 2 & 9 \\ 4 & 1 & 9 \end{bmatrix}$$

and the vector B

$$B = \begin{bmatrix} 5 \\ 13 \\ 7 \end{bmatrix}$$

In Matlab, you enter

```
A = [1,3,7;3,2,9;4,1,9]
```

To solve the system by Cramer's rule, you must calculate some determinants

```
>> det(A);
```

For x1, make a determinant whose first column is the same as vector B and second and third columns are the same as vector A.

```
>> D1 = A; D1(:,1)=B;
```

For x2, make a determinant whose second column is the same as vector B and first and third columns are the same as vector A.

```
>> D2 = A; D1(:,2)=B;
```

For x3, make a determinant whose third column is the same as vector B, first and second columns are the same as vector A.

```
>> D3 = A; D1(:,3)=B;
```

The solution by Cramer's rule is

```
>> X = [ det(D1); det(D2); det(D3);]/det(A)

X =
        -124
        -118
          69
```

The \ and / Division Operators

Matlab uses the \ division symbol to solve matrices representing simultaneous linear equations. There are two situations where the unknown matrix appears on the left or right of the coefficient matrix. Consider the following equation,

$$3x = 9$$

```
>> A=[3];
>> B = [9];
```

In matrix terms, it is the equivalent of $AX = B$ and to solve the equation, the coefficient matrix A should be in the denominator. Thus, using the \ division operator, the unknown matrix could be solved by,

```
>> X=A\B
X =
     3
```

If an unknown appears on the left side, $XA = B$, then use the / division,

```
>> X = B/A

X =
    3
```

In most situations, the number of unknowns equals the number of equations, thus, resulting in a square matrix. A square matrix A is singular if it does not have linearly independent columns. If A is singular, the solution to $AX = B$ either does not exist, or is not unique. The \ operator, A\B, issues a warning if A is nearly singular and raises an error condition if exact singularity is detected.

Polynomial Least Square Fit

The \ operator also performs the least square solution. Consider an experimental result where a quantity y is measured at time value t,

```
>> t=[0 .3 .8 1.1 1.6 2.3];
>> y =[.82 .72 .63 .60 .55 .50]
```

The data can be modeled with a decaying exponential function,

$$y(t) \approx c_1 + c_2 e^{-t}$$

The equation says the vector y should be approximated by a linear combination of two other vectors, one the constant vector containing all ones and the vector with components e-t. The unknown coefficients, c1 and c2 may be computed by doing least square fit. There are six equations and two unknowns, represented by a 6-by-2 matrix.

```
>> E = [ones(size(t)) exp(-t)]
E =
    1.0000    1.0000
    1.0000    0.7400
    1.0000    0.4493
    1.0000    0.3329
    1.0000    0.2019
    1.0000    0.1003
```

The least square solution is obtained by the \ operator.

```
>> c=E\y

c =

        0.4760
        0.3413
```

In other words, the least square fit to the data

$$y(t) \approx 0.4760 + 0.3413e^{-t}$$

We can use the equation to evaluate the model at regularly spaced intervals.

```
>> T = (0:0.1:2.5);
>> Y = [ones(size(T)) exp(-T)]*c;
>> plot(T,Y,'-',t,y,'o')
```

Polynomial Regression

The \ operator can also be used to find the coefficients of a polynomial function that describes the model of the system. For example, the following set of data taken at the time interval t,

```
>> t=[0 .3 .8 1.1 1.6 2.3];
>> y =[.82 .72 .63 .60 .55 .50]
```

The polynomial function that we are trying to model is,

$$y(t) = c_1 + c_2 t + c_2 t^2$$

The unknown coefficients c_1, c_2, and c_3 may be computed by doing the least square fit. There are six equations and three unknowns,

$$\begin{bmatrix} y1 \\ y2 \\ y3 \\ y4 \\ y5 \\ y6 \end{bmatrix} = \begin{bmatrix} 1 & t_1 & t_1^2 \\ 1 & t_2 & t_2^2 \\ 1 & t_3 & t_3^2 \\ 1 & t_4 & t_4^2 \\ 1 & t_5 & t_5^2 \\ 1 & t_6 & t_6^2 \end{bmatrix} \times \begin{bmatrix} c_1 \\ c_2 \\ c_3 \end{bmatrix}$$

```
>> X=[ones(size(t)) t t.^2];
>> c = X\y
c =
        0.5318
        0.9191
       -0.2387
```

Thus, the second order polynomial model of the data is therefore

$$y(t) = 0.5318 + 0.9191t - 0.2387t^2$$

This equation can be used to evaluate the model at regularly spaced intervals.

```
>> T = (0:0.1:2.5);
>> Y = [ones(size(T)) exp(-T)]*c;
>> plot(T,Y,'-',t,y,'o')
```

FLOW CONTROL

Matlab supports the following flow control constructs:

```
if (expression)
..
elseif (expression)
..
else
..
end
```

The `if` and `elseif` evaluate a logical expression and execute a group of statements if the expression is true. The `if` block is terminated until an `elseif`, `else`, or `end` statement is found. The statements followed by the `else` block are executed when the `if` or the `elseif` expression results in false.

The comparison operators are ==, <, >, isequal, isempty, all, and any.

You can check for equality with,

```
>> if A==B
```

Or

```
>> if isequal(A,B)
```

switch and case

```
switch (expression)
case 0
 ..
case 1
 ..
case 2
 ..
otherwise
 ..
end
```

The switch statement evaluates a logical expression and jumps to a case statement that matches the result of the expression. If no match is found, the statement followed by otherwise is executed.

```
while (expression)
 ..
 ..
end
```

The while statement evaluates a logical expression and executes the statements until end is found, if the expression results true,

```
for number of repeats
 ..
end
```

The for loop repeats a group of statements for a fixed, predetermined number of times,

```
>> for a=1:5
       A(a,1)=0
    end
```

```
break
```

The break statement lets you exit early from a for or while loop.

SIGNAL PROCESSING

If you have Signal Processing Toolbox installed you can design filters and plot the graphs and generate waveforms on the command line.

Generating Waveforms

Get one cycle worth of time vector 0 ..2*pi with steps of .001,

```
>> t = 0: .001:2*pi;
```

Plot 1 Hz of a sin function

```
>> plot(sin(t))
```

Plot a 4 Hz sin function

```
>> plot(sin(4*t))
```

Sawtooth Waveform

```
>> t = 0:.001:8pi;
>> plot(sawtooth(t))
```

Chirp Waveform

Get 2 seconds time interval, sampling rate 1 kHz

```
>> t=0:1/1000:2;
```

Get chirp data that starts at 100 Hz and crosses 200 at 1 second

```
>> A = (chirp(t,100,1,150));
```

Spectrogram from the Signal

The spectrogram splits the signal into overlapping segments and displays the frequency contents of each segment.

```
spectgram(A, NFFT, Fs, Window, Noverlap);
```

A: The signal vector
NFFT: Length N FFT
Fs: Sampling frequency
Window: Windows of overlapping segments
Noverlap: Number of samples the section A overlaps

```
spectgram(A, 256, 1000, 256, 250);
```

Sinc Function

The `sinc` (sinx/x) function is the Fourier Transform of the rectangular functions of either time or frequency.

Get a linearly spaced vector length of 100 points

```
>> t=linspace(5,5) ;
>>y=sinc(t);
```

Generate a Unit Impulse Function

Get a time vector column data

```
>> t=(0.001:1)';
>> y=[zeros(size(t))];
>>y(1,1)=1;
```

Generate a Step Function

```
>> t=(0.001:1)';
>> y=[ones(size(t))];
```

Generate a Ramp Function

```
>> t=(0.001:1)';
>> y=t;
```

Convolution

Convolution is a foundation of filtering. The impulse response of a filter multiplied by the input signal is the same as a frequency filter. An averaging filter may be created as,

```
>> h = [1 1 1 1]/4;
>> x = rand(10,1);
>> y=conv(h,x);
```

TRANSFER FUNCTIONS

A system Transfer Function may be represented in several different ways. The most basic is the numerator and denominator polynomial form, providing the input and output relationship as the z-Transform of the difference equation. The other representations are discussed in the following sections.

Zero-Pole-Gain (zpk)

The factored form is the zero-pole-gain form. The `poly` and `root` functions convert between polynomial and zero-pole-gain,

```
>> b = [2 3 4];
>> a = [1 2 3 4];
>>r1 = roots(b);
>>r2 = roots(a);
>> k = b(1)/a(1);
```

State-Space

A system of difference equation may be represented as a set of first order difference equation, with A as the order of the filter, B is a column vector, C is a row vector and D is a scalar,

For example, the difference equation $my'' + cy' + hy = F$ may be written as

$$A = \begin{bmatrix} -c/m, -h/m \\ 1, 0 \end{bmatrix}, \quad B = \begin{bmatrix} 1/m \\ 0 \end{bmatrix}, \quad C = [0,1], \quad D = [0]$$

You can convert a Transfer Function into state-space form with the function `tf2ss`, and the function `ss2tf` converts back into polynomial form.

```
>> num =1;
>> den = [2 2 1];
>> [A B C D] = tf2ss(num,den);
```

Partial Fraction Expansion (Residue Form)

A Transfer Function may be converted into partial fraction expansion form by the function `residuz`,

```
>> b=[-4 8];
>> a = [1 3 4];
>> [r,p,k]=residuz(b,a);
```

Second Order Section (SOS)

A Transfer Function of several orders may be simplified to second order section; the functions `sos2tf` and `tf2sos` perform the conversion.

Lattice Structures

For each pole and zero there are corresponding lattice structure coefficients, also called reflection coefficients of the filter. The magnitude returned provides an easy check for stability. If all the reflection coefficients are less than 1, all the roots are inside the unit circle. You may implement a discrete filter with the lattice coefficients as,

```
>> a = [1 0.6149 0.9899 0 0.0031 -0.0082];
>> k = tf2latc(a);
```

Discrete Fourier Transform

The function `fft` computes the DFT.

```
>> t=(0:.1:1);
>> x = sin(2*pi*t);
>> y=fft(x);
```

FILTER DESIGN

Matlab supports a range of filter design functions. The basic filter function takes a Transfer Function where *b* represents the coefficient of the numerator polynomial and *a* represents the coefficient of the denominator polynomial and performs the convolution with the input *x*; a filter may be implemented with the filter function as,

```
>> b =1;
>> a =[1 -0.9];
>> y = filter(b,a,x);
```

Removing Phase Distortion from a Filter

The IIR filter phase distortion may be removed by the function `filtfilt` that looks into the future values. The filter is also called a noncausal filter.

```
>> b= ones(1,10)/10; % 10 point averaging filter
>> y=filtfilt(b,1,x);
```

Impulse Response

A filter's impulse response may be obtained with the function `impz`,

```
>> imp = [1; zeros(99,1];
>> h = filter(b,a,imp);
```

Frequency Response Bode Plot

The `freqz` function computes frequency response at *n* different points and displays the Bode plot.

```
>> n=100;
>> [b,a]=cheby1(12,0.5,100/500);
>> freqz(b,a,n);
```

Delay

The group delay of the filter is the average delay of the filter as a function of frequency. The function grpdelay computes the group delay for *n* number of points.

```
>> grpdelay(b,a,n);
```

Pole-Zero Analysis

The zplane function plots poles and zeros of a linear system.

```
>> z = -0.5;
>> p = .8*exp(j*2*pi*[-0.4 0.4]);
>> zplane (z,p);
```

IIR Filters

The Transfer Function of the IIR filters have numerator and denominator polynomials as *b* and *a*. The following types of IIR filters are supported.

Chebyshev Type 1 Filters

The Chebyshev type 1 filters have ripples in the pass band. The syntax for type 1 is,

```
cheby1(n,Rp,Wc,'ftype').
```

Wc is the angular cutoff frequency (between 0 and 1, where 1 represents the Nyquist frequency), *n*th order, ripple Rp in dB of peak-to-peak ripple in the passband. The function returns the coefficients in row vectors b, a.

For example, for a fifth order low-pass filter with .1 cutoff frequency, ripple 2dB, enter,

```
>> [b,a]=cheby1(5, 2, .1);
>> freqz(b,a);
```

For a fifth order high-pass filter with .8 cutoff frequency, ripple 2dB, enter,

```
>> [b,a]=cheby1(5, 2, .8);
>> freqz(b,a);
```

For example, for a fifth order band-pass filter with .1 to .2 cutoff frequency, ripple 2dB, enter,

```
>> [b,a]=cheby1(5, 2, [.1 .2]);
>> freqz(b,a);
```

A fifth order band-stop filter with .1 to .2 cutoff frequency, ripple 2dB, enter,

```
>> [b,a]=cheby1(5, 2, [.1 .2], 'stop');
>> freqz(b,a);
```

Chebyshev Type II Filters

The Chebyshev type II filters have ripples in the stop band. The syntax for type II is,

```
cheby2(n,Rp,Wc,'ftype').
```

Butterworth Filters

The syntax for Butterworth is,

```
butter(n,Wc,'ftype').
```

n is the order, Wc is the angular cutoff frequency (between 0 and 1, where 1 represents the Nyquist frequency). You may omit ftype if the filter is not a high-pass or stop-band. The function returns the coefficients in row vectors b,a.

To design a fifth order 20 Hz low-pass filter, for a sampling rate of 100 Hz, enter,

```
>> [b,a] = butter(5, 30/50);
>> y=freqz(b,a);
```

Elliptic Filters

The syntax for an Elliptic filter is,

```
ellip(n,Rp,Rs,Wc,'ftype').
```

n is the order, ripple Rp in dB of peak-to-peak ripple in the pass band, attenuation Rs in dB in the stop-band. Wc is the angular cutoff frequency (between 0 and 1, where 1 represents the Nyquist frequency). You may omit ftype if the filter is not a high-pass or stop-band. The function returns the coefficients in row vectors b,a.

To design a fifth order 200 Hz low-pass filter, ripple in the pass band .2 dB, attenuation of 50dB in stop band and sampling rate of 1000 Hz, enter,

```
>> [b,a] = ellip(5, .2,50,200/500);
>> y=freqz(b,a);
```

The Windowing Method and FIR Filters

The `fir1` function designs FIR filters in the standard range of low-pass, high-pass, band-pass, and band-stop range, whereas `fir2` filter function designs in an arbitrary linear frequency response.

The `sinc(x)` function returns the coefficients corresponding to the impulse response of the basic rectangular filter. There is no denominator polynomial for FIR filters

For example, to obtain 31 coefficients of a low-pass filter with cutoff frequency 0.4pi radians per sample,

```
>> b = 0.4 sinc(0.4(-15:15));
>> freqz(b,1);
```

To obtain a windowed response, compute the window coefficients and multiply with the rectangular coefficients of the impulse response.

The Hamming Window

```
>> b = 0.4 sinc(0.4(-15:15));
>> b =b.*hamming(31)'
>> freqz(b,1);
```

The Von Hann Window

```
>> b = 0.4 sinc(0.4(-15:15));
>> b =b.*hanning(31)'
>> freqz(b,1);
```

Appendix B: Scilab Tutorial

Scilab is similar to Matlab, it is an interactive program operating upon matrices, with graphic capabilities, and it is extensible by user-written functions. The goal of this tutorial is to provide a hands-on introduction to Scilab. It is open source software, freely available to download, installs flawlessly, and you can start using it just a few minutes after downloading.

The following is the link to the Scilab homepage: *http://scilabsoft.inria.fr/*.

The data structures in Scilab include lists, polynomials, rational functions, linear systems, and matrices similar to Matlab, but the syntax is different.

Once you execute Scilab, an interactive shell is open. You can enter the instructions and see the response immediately. To get help on any function, just enter `help` and the function name. The accompanying CD-ROM directory Scilab contains scripts with Scilab instructions for creating figures and example solutions.

The following is a short tutorial on Scilab.

SCILAB BASICS

First, type the following:

```
‡ help plot
```

To see how the figures in Chapter 7, "Digital Filters," are created, change to that directory on the CD-ROM.

```
‡ chdir d:/scilab

‡ exec script7.sci
```

Complex number addition

```
‡ complex=3+2*%i + 4+5*%i
 complex  =

    7. + 7.i
```

Create a sin function of 100 points and plot it,

```
‡ plot(sin(2*%pi*[0:.01:1]))
```

A 20 Hz sine wave, total samples of 100 should have a period of 20/100 = 0.2

```
‡ k=1:100; sinf = sin (2 * %pi * 0.2 * k); plot(sinf);
```

A composite signal of modulated sinusoid at 10 Hz, sampled at .02 sec apart (50Hz sampling), 5 sec total,

```
‡ t = 0:0.02:5;

‡ signal = sin(2 * %pi * 9.7 * t) + sin(2 * %pi * 10.3 * t);

‡ plot2d(t,signal);
```

Polynomial Operations

Create a polynomial that has roots on 1, 4, –5

```
‡  q=poly([1 3 -4],'x')
q  =
12 - 13x + x3
‡
```

Create a polynomial with coefficients $x_2 + 2x + 1$ and see its roots,

```
‡ r=roots(poly([1 2 2],'x','c'))
r  =
! -  .5 +  .5i !
! -  .5 -  .5i !
‡
```

Create a polynomial as a characteristic of a matrix, which is defined by det$[sI-a]$

```
‡ a=[2 3; 4 6]
 a   =
!    2.     3. !
!    4.     6. !

‡ poly(a,'x')
 ans  =
  - 7.105E-15 - 8x2 + x
```

Clean up the zero in the result

```
‡ clean (poly(a,'x'))
 ans  =

  - 8x2 + x

‡
```

The evaluation of roots of characteristics polynomial results in the Eigenvalues of the matrix a.

```
‡ roots (clean(p))
 ans  =

!   0  !
!   8. !
```

Polynomials can be added, subtracted, multiplied, and divided by other polynomials.

```
‡ x=poly(0,'x')
x =
x

‡ q1=3*x+1
q1 =
1 +3x
‡ q2=x^2-2*x+4
q2 =
```

```
          4 -2x2 +x
‡ q2+q1
ans =

5 +x2 +x

‡ q2-q1
ans =ans =
2
3 -5x +x
‡ q2*q1
ans =
       2   3
4 +10x -5x+3x
‡ q2/q1
ans =

          2
4 -2x +x
─────
1 +3x
‡ q2./q1

ans =

          2
4 -2x +x
─────
1 +3x

‡ p=poly([1 2],'x')./poly([3 4 5],'x')
p =
         2
2 -3x+x
─────────
             2  3
-60 +47x-12x +x
```

Evaluation of Polynomials

Polynomials may be evaluated at specific points. The operation is useful for finding Transfer Functions and plotting. Evaluation of polynomials is accomplished using the primitive freq. The syntax of freq is as follows

```
pv=freq(num,den,v)
```

The argument v is a vector of values at which the evaluation is needed.
   Numerator polynomial

```
‡ inval = poly([1 2],'x','c');
```

Denominator polynomial

```
‡ outval = poly([1 2 3],'x','c');
```

Evaluating the input/output at points 0..100;

```
‡ h=freq(inval,outval,[0:100]);
```

The graph of h

```
‡ plot(h)
```

A plot of a 3° polynomial

```
‡ outval = poly([1 2 3],'x','c');

‡ plot(freq(outval,1,[-5:.01:5]));
```

Plotting

A sin function with sampling interval of .02:

```
‡ t = 0:0.02:5;

‡ f1 = sin(2 * %pi * t);

‡ plot(f1);
```

Superimposing another plot:

```
‡ f2 = sin(2 * %pi *1.5* t);

‡ plot(f2);
```

Clear the graph:

```
‡ xclear()
```

Setting boundaries of the plot based on the min and max of the x and y data. (See the help on plot2d for arguments to the plot2d function.)

```
‡ plot2d(t,f1,strf="021");
```

EXAMPLE B.1

Plotting the second curve with different legends,

```
‡ plot2d([t',t'],[f1',f2'],style=[0,1],strf="021");
```

EXAMPLE B.2

Plotting with x and y coordinates from a rectangle arguments

```
‡ plot2d([t',t'],[f1',f2'],style=[0,1],strf="051",rect=[-2,-2,8,2]);
```

Multiple graphs on the same window,

```
‡ s1 = f1* .1+ rand(f1) * 2;

‡ s2 = f2 * .3+ rand(f2) * 2;

‡ xsetech ([0,0,1,1/4])
                            // defines first subwindow
‡ plot2d(t,s1,1,"050"," ",[-2,-2,8,2]);
                            // plot first trace, no boundary
‡ xsetech ([0,1/4,1,1/4])
                            // defines second subwindow
‡ plot2d(t,s2,1,"050"," ",[-2,-2,8,2]);
                            // plot second trace, no boundary
```

The following primitives affect the graph output:

xset: Set values of the graphics context
xselect: Raise the current graphics window
xclear: Clear a graphics window
xbasr: Redraw a graphics window
xbasc: Clear a graphics window and erase the associated recorded graphics
xdel: Delete a graphics window

Draw arcs with the xarc function:

```
xarc(-1.5,1.5,3,3,0,360*64);
xarc(1,1,.2,.2,0,360*64);
xarc(-1,1,2,2,0,90*64);
```

THE TRANSFER FUNCTION

The function freq may be used for the evaluation of frequency response of filters and system transfer functions. For example, a discrete filter can be evaluated on the unit circle in the z plane as follows.

First we create an FIR filter with rectangular impulse response

```
‡ h=[1:21];
```

Create a polynomial representing the z-Transform of the filter

```
‡ hz=poly(h,'z','c');
```

Evaluating the polynomial at the points $\exp(2\pi i n)$ for $n = 0, 1, \ldots, 20$, which amounts to evaluating the z-Transform on the unit circle at 21 equally spaced points in the range of angles $[0, \pi]$.

```
‡ f=(0:.01:1);
```

```
‡ hf=freq(hz,1,exp(%pi*%i*f));
```

Display the result:

```
‡ plot(abs(hf'))
```

CONVOLUTION

Convolution of time-based signals converts them into frequency-based signals and vice versa. The command used by Scilab to design the convolution is `convol`.

```
‡ a = [1 2 3]
  a =

!   1.    2.    3. !

‡ b = [1 0 0 0 0 1 0 0 0]
  b =

!   1.    0.    0.    0.    0.    1.    0.    0.    0. !

‡ c = convol(a,b)
  c =

!   1.    2.    3.    0.    0.    1.    2.    3.    0.    0.    0. !
```

Convolution of an impulse response with the input signal acts as a filter.

High-pass filter: eliminating the slow components could be accomplished by subtracting each value from the previous one.

EXAMPLE B.3

```
‡ t = 0:0.02:5;

‡ f1 = sin(2 * %pi * t);

‡ s1 = f1+ rand(f1) * 2;

‡ s1hp = convol([1,-1],s1);

‡ xsetech ([0,0,1,1/3])

‡ plot(s1);

‡ xsetech ([0,1/3,1,1/3])

‡ plot(s1hp);
```

Low-pass filter: eliminating the high-frequency components could be accomplished by adding the current value to the previous one.

```
‡ s1lp = convol([1,1],s1);

‡ xsetech ([0,2/3,1,1/3])

‡ plot(s1lp);
```

FAST FOURIER TRANSFORMS

The Scilab primitive for the Fast Fourier Transform is fft. In the following example, data from a cosine function is passed to the fft function,

```
‡ x=0:63;y=cos(2*%pi*x/16);

‡ yf=fft(y,-1);

‡ plot(x,real(yf)');
```

CORRELATION

Correlation measures the degree at which both signals are related. The command to perform correlation is corr. It can be used with one or two vectors. If we call corr with only one vector, we get the correlation of the vector with itself (autocorrelation). If we use two vectors, we get the correlation between them (crosscorrelation).

EXAMPLE B.4

We will simulate a sine wave at 10 Hz, using a 10 second signal sampled at 100 Hz.

```
‡ t = 0:.01:10;
‡ s1 = sin(2 * %pi * 10 * t);
```

Create a Gaussian noise of the same length, also called *white noise* as it contains all frequencies. Calling the function rand with the option normal creates a distribution of values with a mean of 0 and a standard deviation of 1. We multiply by 2 to get a normal distribution of values with a mean of 0 and a standard deviation of 2.

```
‡ rand('normal')
‡ rand('seed',0)
‡ noise = 2 * rand(s1);
‡ s1noise = s1 + noise;
```

Perform autocorrelation

```
‡ autocorrs1 = corr (s1,1000);
```

Perform cross correlation

```
‡ crosscorrs1 = corr (s1,s1noise,1000);
```

Combine the four graphs in one window,

```
‡ xsetech([0,0,1,1/4])
‡ plot2d(t(1:500),s1(1:500));
‡ xsetech([0,1/4,1,1/4])
‡ plot2d(t(1:500),s1noise(1:500));
‡ xsetech([0,1/2,1,1/4])
‡ plot2d(t(1:500),autocorrs1(1:500))
‡ xsetech([0,3/4,1,1/4])
‡ plot2d(t(1:500),crosscorrs1(1:500))
```

IIR FILTERS

The Scilab function for creating IIR filters has the following format,

```
‡ [hz]=iir(n,ftype,fdesign,frq,delta)
```

The argument n is the filter order, which must be a positive integer. The argument ftype is the filter type and can take values,

lp: For a low-pass filter

hp: For a high-pass filter

bp: For a band-pass filter

sb: For a stop-band filter

The argument `fdesign` indicates the type of analog filter design technique is to be used to design the filter. The value of `fdesign` can be,

butt: For a Butterworth filter design

cheb1: For a Chebyshev filter design of the first type

cheb2: For a Chebyshev filter design of the second type

ellip: For an elliptic filter design

The argument `frq` is a two-vector, which contains cutoff frequencies of the desired filter. For low-pass and high-pass filters, only the first element of this vector is used. The first element indicates the cutoff frequency of the desired filter. The second element of this vector is used for band-pass and stop-band filters. This second element is the upper-band edge of the band-pass or stop-band filter, whereas the first element of the vector is the lower-band edge.

The argument `delta` is a two-vector of ripple values. In the case of the Butterworth filter, `delta` is not used. For Chebyshev filters of the first type, only the first element of this vector is used and it serves as the value of the ripple in the pass band. Consequently, the magnitude of a Chebyshev filter of the first type ripples between 1 and 1-`delta(1)` in the pass band. For a Chebyshev filter of the second type, only the second element of `delta` is used. This value of `delta` is the ripple in the stop band of the filter. Consequently, the magnitude of a Chebyshev filter of the second type ripples between 0 and `delta(2)` in the stop band. Finally, for the elliptic filter, both the values of the first and second elements of the vector `delta` are used and they are the ripple errors in the pass and stop bands, respectively.

The output of the function `hz` is a rational polynomial giving the coefficients of the desired filter.

Example B.6 describes the use of `iir` function.

EXAMPLE B.5

Design a low-pass Chebyshev filter of the first type, cutoff frequency of .1, filter order 4, and the ripple in the pass band of .05.

```
‡ hz=iir(4,'lp','cheb1',[.1 0],[.05 0])
```

Use the bode plot function to plot the magnitude and phase response curve in logarithmic units.

```
‡ bode(hz)
```

The upper frame of the bode plot is the magnitude response of the filter, and the lower frame is the modification in phase that the system introduces.

The frequencies are expressed as powers of 10 and span from 0.001 to 1 (the x axis is equivalent to 0.001, 0.01, 0.1, and 1). Frequencies above 0.5 do not exist. If sampled at 200 Hz, the values in the x-axis correspond to frequencies of 0.2 (0.001 * 200), 2 (0.01 * 200), and 20 (0.1 * 200) Hz. The magnitude is indicated in decibels.

The dB or decibel value in decimal is computed as,

```
‡ decval = [1 0.1 0.001 0.001]    //decimal values
  decval =

!   1.    0.1    0.001    0.001 !

‡ dbval = 20 * log10 (decval)   //db values
  dbval =

!   0.   - 20.   - 60.   - 60.  !
```

And the decimal values from dB are obtained by,

```
‡ dbval = [0 -3 -6 -20]          //db values
  dbvalues =

!   0.   - 3.   - 6.   - 20.  !

‡ decval = 10 ^ (dbval/20)       //decimal values
  decval =

!   1.    0.7079458    0.5011872    0.1 !
```

Transfer Function Magnitude

The Scilab function cheb1mag may be used to obtain the magnitude of the filter Transfer Function for plotting purposes

```
‡ n=13;epsilon=0.2;omegac=3;sample=0:0.05:10;

‡ h=cheb1mag(n,omegac,epsilon,sample);

‡ plot(sample,h,'frequencies','magnitude')
```

A plot of the Transfer Function magnitude describes how the system behaves at different input frequencies. The Scilab command `frmag` returns a vector with the output of the system when the input varies in frequency. The format of the `frmag` function is,

```
[famp, f] = frmag(numer(fltr), denom(fltr), 256);
```

The first two inputs (`numer` and `denom`) are the numerator and the denominator polynomials of the filter, and the third input is an integer indicating the length of the array that will be returned. The two vectors are the frequencies `f` and the amplitudes of the output (`famp`). The `frmag` returns frequencies in the range of 0 to 0.5. To evaluate the output at the sampling frequency used, multiply the `f` by the sampling frequency used, in this case 200. Continuing with the previous example, the following instructions plot the magnitude response of the filter of Example B.6.

```
‡ xsetech([0,0,1,1])

‡ xbasc()

‡ plot2d(f*200,famp)
```

Poles and Gain

The poles and gain of Chebyshev filters may be obtained with the function `zpch1`,

```
‡ n=13;epsilon=0.2;omegac=3;sample=0:0.05:10;

‡ [p,gain]=zpch1(n,epsilon,omegac);

‡ tr_fct=poly(gain,'s','coef')/real(poly(p,'s'))

‡ [pols,gn]=zpch1(3,.22942,2);

‡ gn
gn =
8.7176358
```

```
-->pols'
 ans =
 ! - 0.7915862 - 2.2090329i !
 ! - 1.5831724 - 1.562D-16i !
 ! - 0.7915862 + 2.2090329i !

-->hs=gn/real(poly(pols,'s'))

-->fr=0:.05:3*%pi;

-->hsm=abs(freq(hs(2),hs(3),%i*fr));

-->plot(fr,hsm)
```

To see the filter in action, we will pass sample data (containing two frequencies, one below the filter cutoff frequency and the other one above the cutoff frequency) through the filter and verify the result to see the low-frequency signal vanish.

Sample the same signal at 200 Hz, (sampling period of 0.005). The filter cutoff point of .1 indicates .1*200 = 20Hz as the cutoff frequencies. We expect all the frequencies below 20 Hz to be filtered out with the filter of Example B.5.

Begin with the sampling time base, taking four cycles worth of data

```
-->t=.01:.005:4;
```

A signal of frequency 20 Hz is equivalent to (20/200) =.1 of the sampling frequency,

```
-->sig1=sin(2*%pi*20*t);
```

A signal of frequency 1 Hz is equivalent of (1/200) = .005 of the sampling frequency,

```
-->sig2=sin(2*%pi*1*t);

-->sig=sig1+sig2;
```

Create a Chebyshev low-pass filter of cutoff frequency of .05 and maximum ripple of .05,

```
-->hz=iir(4,'lp','cheb1',[.05 0],[.05 0])
```

Apply the filter using the Scilab function flts,

```
‡ out1 = flts(sig,hz);
```

Similarly, we can eliminate the low-frequency component from the sig data by passing the data through a high-pass filter of cutoff frequency .1 and maximum ripple of .05,

```
‡ hz=iir(4,'hp','cheb1',[.05 0],[.05 0])
```

Apply the filter using the Scilab function flts,

```
‡ out2 = flts(sig,hz);
```

Butterworth Filters

Scilab has a special function (buttmag) for obtaining Butterworth filter squared magnitude response. The syntax is,

```
[h]=buttmag(order,omegac,sample)
```

The order is the filter order, omegac is the cutoff frequency, and the sample is the vector of frequency where buttmag is evaluated. The following example demonstrates a Butterworth filter with the cutoff frequency of 200 Hz and order 5

```
‡ h=buttmag(5,200,1:1000);

‡ mag=20*log(h)'/log(10);

‡ plot2d((1:1000)',mag,[1],"011"," ",[0,-180,1000,20]),
```

The poles and zeros of the filter may be obtained with the function zpbutt,

```
‡ n = 5;

‡ omegac = 2;

‡ [poles,gain]=zpbutt(n,omegac);
```

Obtain the Transfer Function polynomial with,

```
‡ h=poly(gain,'s','coeff')/real(poly(pols,'s'))
```

Window FIR Filters

The FIR filters generally have a uniform delay in their phase response. The Scilab function `wfir` designs the four basic FIR filters: the low-pass, high-pass, band-pass, and band-stop filters. We only have to specify the low-pass filter impulse response, and the other three may be obtained by simple relations. The general form of the `wfir` filters is given as,

```
[wft,wfm,fr]=wfir(ftype,forder,cfreq,wtype,fpar)
```

The `ftype` is the filter type: `lp`, `hp`, `bp`, `sb`. The `forder` is a positive integer giving the order of the desired filter.

The `cfreq` is a two-vector argument, but the low-pass and high-pass filters use only one, whereas the band-pass and band-stop filters require two.

Some other terms are:

wtype: The window type,
re: Rectangular
tr: Triangular
hm: Hamming
kr: Kaiser
ch: Chebyshev

The `fpar` parameter is specific to the window type. In the case of a Kaiser window, the first element of `fpar` indicates the relative trade-off between the main lobe of the window frequency response and the side-lobe height, and this must be a positive integer. For the case of the Chebyshev window, one can specify either the width of the window's main lobe or the height of the window side lobes. The first element of `fpar` indicates the side-lobe height and must take a value in the range 0 < argument < 1, the second element gives the main-lobe width and must take a value in the range 0 < argument < 0.5. If you don't want to specify a value, leave it as a negative number. Thus, `fpar=[.01,-1]` means that the Chebyshev window will have side lobes of height .01 and the main-lobe width is left unspecified.

The return parameters are,

wft: Time domain filter coefficients
wfm: Frequency domain filter response on the filter grid `fr`
fr: Frequency grid

Appendix C
Digital Filter Applications

There is hardly any field in engineering that does not benefit from the techniques of digital signal processing. In the realm of image processing, audio processing, digital communication, neural networks, pattern recognition, biomedical applications, astronomy, and in countless other applications we find digital signal processing playing an important role of processing data that produces meaningful results.

In this Appendix, three digital filtering applications are presented: image processing, audio processing, and filter testing. The image processing application features edge enhancement, smoothing, sharpening, inverting, and edge detection operation on a digital image stored as a bitmap file. The audio processing application detects the presence of a specific frequency in a stream of audio data acquired through devices such as the sound card in a PC. The frequency detection is a common application in telephony, where dial tones and dialing digits are transmitted as predefined frequencies over the telephone line. Transfer of data over the telephone requires digital filter applications that detect the presence of these frequencies in the audio signals. The filter test program is a simple tool to analyze poles and zeros of IIR filters.

The applications are being developed using the Perl/Tk modules. The intent is to present short and simple programs that demonstrate the use of digital filtering.

IMAGE PROCESSING

Images are captured as points in space, each with a certain level of intensity with reference to certain spatial coordinates and stored in computer memory as numbers reflecting the relative strength. Storing images in color requires more information than simple grayscale images, but the fundamentals of processing is the same for both, color as well as black and white with shades of gray.

The number of pixels in an image depends upon the rendering process. Different file formats have different structures. Image files are usually stored with header information followed by image data. The header carries the necessary parsing information about the image such as the number of rows, columns, colors, etc.

The application presented in this chapter is for processing grayscale images only, but the technique is equally applicable to color images. If you think of a grayscale image as a two-dimensional matrix representing a single layer, then a color image is a picture with three or four different layers of intensity information. It is sufficient to represent a grayscale pixel using a single byte (256 levels). But the color images require three or four bytes, each red, green, or blue, or cyan, magenta, yellow, and black.

Image processing functions may be grouped into three separate categories: point, neighborhood, and morphological operations. We briefly discuss the objective of each operation, followed by software examples with a graphical user interface that highlights the effect of the operation.

Point Operations

Point operations are functions that are performed on each pixel of an image, independent of all other pixels in the image. The operations include inverting the pixel intensities to produce a negative image, increasing or decreasing the contrast, etc. Point operations can easily be implemented using a lookup table, as there is a one-to-one correspondence of the original image to the translated image.

Image Inverting

Each byte of the grayscale image is an intensity value between 0 and 255, 0 being the lightest shade and 255, the darkest. Making a negative of an image is simply complementing the pixel value by subtracting 255 from the current pixel value. Figure C.1 shows the graphical user interface of the example software. Running the application program displays a bitmap image (see Appendix A, "Matlab Tutorial," for the description of the software). Selecting the option Invert and pressing the button Run displays the modified image that is being displayed on the right side of the frame window.

Increasing or Decreasing the Contrast

Increasing the contrast can be described as scaling the pixel to a higher value, whereas decreasing the contrast is scaling to a lower value. If x is the current pixel intensity value, then multiplying it by 1.5 increases the contrast and multiplying by 0.5 decreases the contrast (if the result is greater than 255, then the value is restored to 255). Figure C.2 shows the effect of the contrast operation when you select Contrast and press Run.

Appendix C: Digital Filter Applications **353**

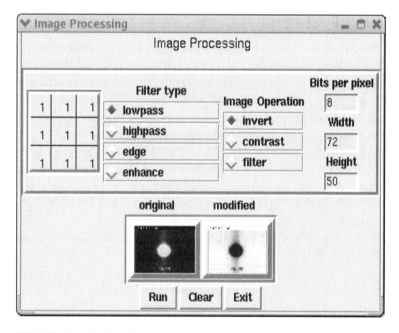

FIGURE C.1 Application program showing image inversion.

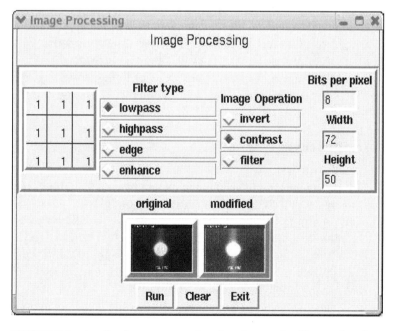

FIGURE C.2 Application program showing the result of increasing the contrast.

Neighborhood Operations

The neighborhood operations usually involve operation on a group of pixels in the neighborhood to produce a new pixel value. The process is essentially a filtering operation. Sharpening, smoothing, and edge enhancement fall into this category. Each operation can be thought of as passing the image data through a high-pass, low-pass, or band-pass filter. A bitmap image with a lot of rapidly changing objects in it may be considered as an image with high-frequency components. These images have sharp edges that can be smoothed by averaging the pixel values in the neighborhood to produce new pixels. It is as if we are passing the image data through a low-pass filter. Similarly, an image in which edges are blended may be considered an image with a lot of low-frequency components. Passing the data through a high-pass filter may sharpen such images. In a two-dimensional image, the filter coefficients are also two-dimensional, as the neighborhood includes the pixels in the immediate vicinity of rows and columns. The two-dimensional filters are also called Mask filters, as if a mask is being passed over each pixel, effectively changing the characteristics of the pixel underneath the mask.

Mask Filters

The Mask filters are usually represented as a window of 3 × 3 matrices, each element of the matrix represents the weight given to the corresponding pixel in the neighborhood. The weight is the filter coefficients, and the coefficient values are chosen to highlight a specific effect. For example, the Mask filter in Figure C.3 gives equal weight to all pixels in the neighborhood, hence producing an averaging effect, whereas the Mask filter in Figure C.4 gives more weight to the pixel in the middle compared to the one surrounding, essentially producing a discriminating effect that enhances the magnitude of the center pixel relative to the one in its surroundings, in effect enhancing the contrast.

Low-Pass and High-Pass Filters

A high-contrast image such as a white object in a dark background has high-frequency components. The pixel value changes from high to low very quickly. Removing sudden changes in the intensity from pixel to pixel smoothes an image (the Mask of Figure C.3 has such an effect); such Mask filters are low-pass filters. With the same token, if the color changes slowly such as a sky in the sunset, that image has a low frequency and low contrast; enlarging the different intensity levels between pixels enhances the contrast. The Mask in Figure C.4 is the high-pass filter that has the effect of enhancing the contrast.

Digital filtering of image data is usually performed after the data has already been acquired, and that makes the filter design an easier process. We do not need to resort to causal filters, since the data from the future is also available for processing. The transformed image is obtained by convolving the image data with the Mask filter, as explained in the next section.

Convolution

The image data and the Mask filter are a two-dimensional matrix. The Mask filter is usually a 3 × 3 matrix, whereas, the number of rows and columns in image data are obtained from the header information of the image file. The convolution of image data with the Mask filter is carried out by multiplying each element in the mask with the corresponding pixel of the image. The result of all the multiplication is added and then divided by the total weight of the mask (which is the sum of the weights of the individual elements). The final value is the value of the transformed pixel. The process is repeated for each image pixel, but special consideration is given to the pixels on the edges, as there is no image data in their immediate periphery. You can either ignore the edge pixels or create hypothetical row and column pixels with 0 values.

The application program imageFilter.pl is provided with a generalized convolution algorithm (see Appendix A for details) with predefined Mask filters for different image transformation operations described next.

Low-Pass Filtering

Figure C.3 is the GUI of the application program imageFilter.pl, displaying the 3 × 3 matrix of the low-pass filter. The filter elements all have equal weights of 1, indicating a low-pass filter kernel. Select Filter from the image operation choice, select Lowpass from the window filter type, and press the Run button. The modified smoother image is displayed in the window modified.

High-Pass Filtering

Continuing with the GUI of the application program imageFilter.pl, select Filter from the image operation choices, select Highpass from the window filter type and press the Run button. The Mask filter now displays the 3 × 3 matrix of the high-pass filter. Notice the center of the kernel carries a weight of 9 compared to surrounding elements that carry a weight of -1. The net effect is a high-pass filter action as explained in the previous section. The modified sharper image is displayed in the window modified.

Edge Enhancement

The filter kernel for edge enhancement is shown in Figure C.5. Run the application imageFilter.pl. Select Filter from the image operation, select Enhance from the window filter type, and press the Run button. The Mask filter now displays the 3 × 3 matrix of the edge-enhancement filter. Notice the center of the kernel carries a

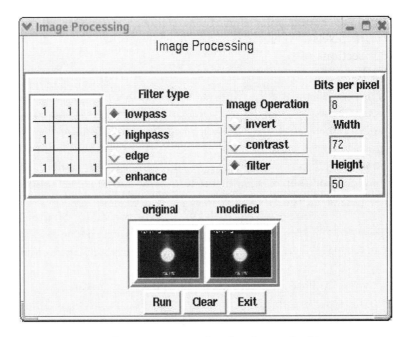

FIGURE C.3 Application program showing the result of low-pass filtering.

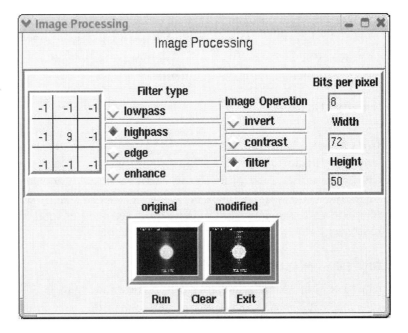

FIGURE C.4 Application program showing the result of high-pass filtering.

weight of 4 compared to the surrounding elements that carry a weight of −1. The corner pixels do not carry any weight. The net effect is the sharpening of the different sections of the image boundaries displayed in the modified window of the GUI.

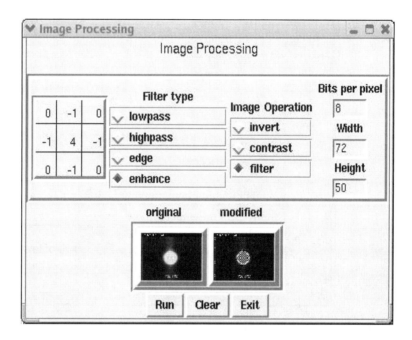

FIGURE C.5 Application program showing the effect of edge enhancement.

MORPHOLOGICAL OPERATIONS

The morphological operations help identify objects and boundaries within an image, such as outlining the different forms and structures in an image. The edge detection is the morphological operation in which the geometry of objects within an image is enhanced.

Edge Detection

The filter kernel for edge detection is shown in Figure C.6. Notice the similarity between high-pass filter kernels and edge detection filter kernels. The center element makes the total weight 0, suggesting a very high threshold value. Though we cannot divide by 0, the threshold is handled by converting all pixels above 125 to 255, and all pixels less than 100 a 0 value. The image is inverted to show black on a white background.

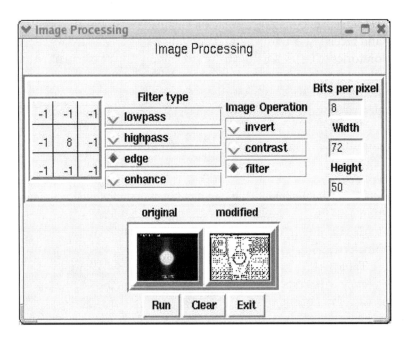

FIGURE C.6 Application program showing the effect of the edge detection kernel.

Run the application imageFilter.pl. Select Filter from the image operation, select Edge from the window filter type, and press the Run button. The Mask filter now displays the 3 × 3 matrix of the edge detection filter. The net effect is outlining of the boundaries of the image and displaying it in the modified window of the GUI.

AUDIO PROCESSING

Audio processing applications, including voice recognition, data compression, voice synthesis, and digital telephony, are candidates of digital signal processing applications, as today's CPUs can easily meet the sampling and computational requirements for the DSP algorithms. The frequency detection is a common application in telephony, where dial tones and dialing digits are transmitted as predefined frequencies commonly known as DTMF or Digital Tone Multi-Frequency signals. Next, we present a digital filter application that detects the presence of these frequencies in the audio signals.

DTMF Signal Detection

DTMF is the signaling mechanism adapted by telephone companies for transmitting and receiving dial tones and dialing digits. The standard was developed by Bell Laboratories to be used in touch-tone telephones and voicemail systems. Each telephone digit corresponds to a high and a low tone that are transmitted simultaneously. Four frequencies in both the high and low group give 16 possible combinations. Table C.1 describes the digits and the corresponding frequencies of DTMF signals.

Performing Fast Fourier Transform would identify all the frequencies in the incoming audio signals, but if the job is to detect few discrete sinusoids such as DTMF signals, it is easier to implement an IIR filter whose magnitude determines the presence of the specific frequency in the audio signal.

In DTMF detection, we only need to identify eight frequencies. The specification asks for a minimum duration of 40 msec. A sampling rate of 8000 is sufficient to detect the highest frequency 1633 Hz specified by the standard. If we choose a block of 256 data, the frequencies are normalized with the sampling rate as shown in Table C.1. (The floating-point k in Table C.1 is the fraction of the sampling frequency with respect to the block size of 256.) We must pick an integer value as the coefficients of the IIR filter, also shown in Table C.1. The output of the filter still needs to be compared against a threshold for detecting the presence of the specific frequency. The Goertzel algorithm is an example of such an implementation as explained in the next section.

TABLE C.1 The DTMF Signal Frequencies and the Corresponding Normalized Frequencies Using 256 Sample Points

Tone in Hz	Floating point k	Integer k	% Error
697	22.304	22	1.3%
770	24.64	25	1.46%
852	27.264	27	0.96%
941	30.112	30	0.37%
1209	38.688	39	0.8%
1336	42.752	43	0.58%
1477	47.264	47	0.56%
1633	52.256	52	0.489%
Dial tone(350)	1.6	2	25%

Goertzel Algorithm

The Goertzel filter is a second order IIR band-pass filter with an extremely narrow bandwidth and a very high gain. But the important point is that after passing a block of data through the filter, a two point discrete Fourier Transform is performed over the last two points of the filter output to identify the presence of the specific frequency. Measure the signal energy before and after the filter and compare the two. If the energies are similar, the input signal is in the pass band, if the output energy is significantly lower than the input, the signal is outside the pass band.

The second order Goertzel filter Transfer Function is specified in Equation C.1.

$$H(z) = \frac{Y(z)}{X(z)} = \frac{1}{1 - 2r\cos\theta z^{-1} + r^2 z^{-2}} \tag{C.1}$$

$$\theta = 2\pi \frac{f_c}{f_s} N$$

The value of $r = 1$ in Equation C.1 makes it a very narrow bandwidth, high gain, band-pass filter. The angle θ is the angular frequency in radians providing the fraction of the Nyquist frequency where the maximum gain is. For example, detecting the presence of a 697 Hz signal in a block of 256 data obtained at 8000 Hz, the coefficient is determined as,

$$N = 256$$
$$f_c = 697$$
$$f_s = 8000$$
$$r = 1$$
$$2r\cos\theta = 2\pi 0.86 = 0.53$$

The filter for 697 Hz having the Transfer Function of the Equation C.1 may be realized with Equation C.2.

$$y[n] = x[n] + 0.53 y[n-1] - y[n-2] \tag{C.2}$$

If your computer is equipped with an audio system (a speaker and a microphone), you can run the example software audio.pl (see Appendix A for details). The program reads the audio signals from the input device (the microphone) and performs the filtering action as specified by Equation C.2. The output of the filter (the magnitude of the DFT) is shown as a vertical bar on the GUI (see Figure C.7). The software is designed using the Perl/Tk module, and the listing audio.pl is provided in Appendix A.

Upon execution of the software, the input device (the microphone) is opened for read. The user may select a specific frequency through the entry dialog box frequency, and by pressing the Run button the software would execute in a continuous loop acquiring a 256-byte block of data from the audio device and perform the convolution of the filter response with the input data. A two-point DFT is performed over the last two bytes of the output of the filter and the magnitude is shown as a vertical bar on the GUI, the loop is repeated until Stop is pressed. In real life, the magnitude should be compared against a threshold, as well as identify, if there is a harmonics of the speech that is giving a false signal. Since DTMF signals do not have second harmonics, we could compute DFT coefficients to determine the second harmonics to detect the presence of speech.

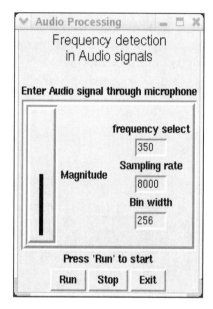

FIGURE C.7 Application program audio.pl showing frequency detection.

FILTERTEST.PL: TESTING THE FILTER DESIGN

One way to test a filter design is to put it through a sweep frequency and view its outcome through a plot of the frequency response and check for the acceptability of the performance. Displaying the outcome as a graph is a must. The filterTest.pl offers a simple tool to analyze the frequency and phase response of the IIR filters and displays the time sequence generated by the filter output. The software is based on a free plotting software (xmgarce) and the Perl/Tk module. xmgrace is a powerful graphic display program with built-in FFT routines that can be easily incorporated into any existing software with a simple pipe interface, and can be analyzed by the listing of the filterTest.pl program. (The listing and the installation procedure are provided in the directory applications of the accompanying CD-ROM.)

You can download xmgrace from the Grace home page: *http://plasma-gate. weizmann.ac.il/Grace/*.

The Perl/Tk module may be downloaded from the Perl home page: *http://www. perl.com*.

Executing the program filterTest.pl opens up a graphical user interface as shown in Figure C.8. The user may enter up to two poles and two zero vectors in polar coordinates and enter up to six different input test frequencies to generate test input function. For Example 7.1, you may enter 0.48 for pole0 magnitude and select 10 Hz, 100 Hz, 200 Hz, 300 Hz, and 400 Hz as input frequencies. Pressing Test will display six different plots: a) input function in time (obtained by combining all the input frequencies), b) plot of "FFT of Input," c) filter response in time (plot section "Output Function"), d) plot of the "FFT of Output," e) the frequency response, and f) the phase response, as shown in Figure C.9. The filter action is clearly displayed by the drop in the magnitude of the frequencies as can be seen from the "FFT of Output" plot.

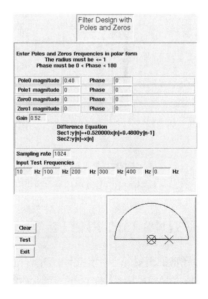

FIGURE C.8 The graphical user interface of the application program filterTest.pl.

The range of the vector r is between 0 and 1, and the phase angle is between 0° and 180°. The pole and zero vectors are real if the phase is 0 or 180; otherwise, it is a complex conjugate vector. For test signals, you can select up to six input frequencies of equal magnitude, you can specify sampling rates varying between 2 and 65536 points. The program also displays the location of the poles and zeros with respect to the origin of the semicircle. A cross indicates the pole-frequency vector, and a circle indicates the zero-frequency vector, inside or right on the edge of the big semicircle. The program displays the following six diagrams,

- Time response of the input data points
- FFT of the input data
- Time response of filter output for one complete cycle
- FFT of output data
- Frequency response of the filter
- Phase response of the filter

Press Test to view the different plots (see Figure C.8).

The following helpful hints will give you some intuitive feelings about the effect of different values of r and the phase angle θ on the filter response.

- If the roots of the poles are real and negative, the filter is a high-pass filter; the peak value is at 180 degree,

$$z + r = 0, z = -r$$

Select input frequencies 1,100,200,300,400,500. Enter 0.8 for the value of the Pole0 radius; enter 1 for Zero0 radius, 180 for the Pole0 phase, 180 degrees for the Zero0, and press Test. You should see frequencies below the 80 percent bandwidth are suppressed.
- If the roots of the poles are real and positive, the filter is a low-pass filter, the peak value is at 0 degrees,

$$z - r = 0, z = r$$

Change the value of pole0 and Zero0 phase angle from 180 to 0; press Clear, press Test. The filter is changed from high-pass to low-pass filter.
- A complex pole as well as a complex zero creates a short band-pass filter. The peak is at the phase angle θ of the complex vector.
- Change the value of pole0 and Zero0 phase angle to 30; press Clear, press Test. The filter is changed to a narrow band-pass filter.
- Increase the bandwidth by reducing the r vector length.
- The same combination of complex pole and zeros creates a band-stop filter, if the zero vector magnitude is greater than the pole vector magnitude. The notch occurs at the phase angle θ.

- Continuing from the previous example, change the value of pole0 magnitude from 0.8 to 1 and Zero0 magnitude from 1 to 0.8; press Clear, press Test. The filter is changed to a narrow band-stop filter.
- The depth of the notch depends upon the difference between the radius of the pole and zero vectors. Increase the bandwidth by reducing the r vector length.
- The response magnitude peak of the pole corresponds to the value $1 - r$ (the minimum length of the pole).
- The 3dB point is where the magnitude falls to $1/\sqrt{2}$ of its peak value determines the bandwidth of the pass band. The bandwidth is approximately two times the vector length $1 - r$.

You can design simple filters by selecting pole and zero locations and viewing the response, but a precision filter requires other properties, such as a steeper roll off of the pass-band region. Ideally, one would like to establish a cutoff frequency and expect a filter to preserve the pass-band range while sharply rejecting the stop band. But such filters are not physically realizable, as they require infinite computations. Still, design methods are available to improve upon the roll-off rate and achieve a good enough response that closely matches the ideal response.

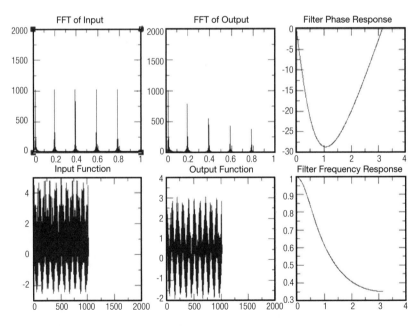

FIGURE C.9 The output generated by the application program filterTest.pl for the parameters of Example 7.1. Six different plots are displayed, including a) Input Function, b) "FFT of Input," c) Output Function, d) "FFT of Output," e) Frequency Response, and f) The Phase Response.

Appendix D: About the CD-ROM

The software has been tested on Linux platform running RedHat Linux 9.0.

CD ORGANIZATION

The following directories and files exist on the CD-ROM:

- ThirdPartySoftware/
  - scilab-2.7.bin.linux-i686.tar.gz
  - grace-5.1.14.tar.gz
  - Tk-804.027.tar.gz
- Applications/
  - IMG/beam.bmp
  - IMG/filterIMG.pl
  - IMG/modified.bmp
  - IMG/README.txt
- Applications/
  - IIR/filterTest.c
  - IIR/filterTest.o
  - IIR/filterTest.pl
  - IIR/Makefile
- Applications/
  - WAV/wav.pl
- matlab/m/
  - fig*.m
- scilab/
  - fig*.sce
  - script*.sci

Scilab is an interactive program for solving math problems similar to Matlab; see Appendix B, "Scilab Tutorial," for details. Grace is a graphics library required by the application program filterTest.pl.

The perl Tk module is required by the application perl scripts.

The three application programs filterIMG.pl, filterTest, and wav.pl demonstrate the techniques being described in the book; see Appendix C, "Digital Filter Applications," for details.

The graphs in this book were developed with the help of Matlab and Scilab scripts. These scripts are provided in the Matlab and Scilab directories. The scripts are named after the figure they generate.

You can execute these scripts from the interactive shells of the Scilab and Matlab.

DOWNLOAD THE LATEST

These items are available on the CD-ROM in the ThirdPartySoftware directory.
You may also download the latest versions:

- Download the latest Scilab from the following Web site: *http://scilabsoft.inria.fr/*
- Download Xmgrace from the following site: *http://plasma-gate.weizmann.ac.il/Grace*
- Download perl/Tk from the following site: *http://www.perl.com/CPAN/*
  - Click "perl modules"
  - Click "all modules"
  - Click to download "Tk-804.027.tar.gz"

INSTALLATION

Installing Scilab Software

- CD to the directory where you want Scilab to reside
- untar the scilab-2.7.bin.linux-i686.tar.gz using the following command:
  `tar -xvzf scilab-2.7.bin.linux-i686.tar.gz`
- CD to the directory scilab-2.7 and do the make utility

This would place the executable Scilab into the scilab-2.7/bin directory

INSTALLING GRACE

The recommended directory to install grace is /usr/local/src/grace-5.1.14.

- CD to the directory /usr/local/src
- untar the grace-5.1.14.tar.gz using the following command:
  tar -xvzf grace-5.1.14.tar.gz
- CD to the directory grace-5.1.14/doc and follow the instructions in the user-guides.html

Copy the grace executable /usr/local/src/grace-5.1.14/src/xmgrace into somewhere in the path such as /usr/local/bin so that it could be launched from anywhere.

The application program filterTest.c includes the "grace_np/grace_np.h" header file and is being linked with the grace library "grace_np/libgrace_np.a." The default directory for xmgrace is assumed to be /usr/local/src/grace-5.1.14. If you have installed grace in any other directory then you must change the "$TOP" variable in the Application/IIR/Makefile to reflect the new directory location.

INSTALLING TK

Untar the Tk-804.027.tar.gz file and follow the instructions in the INSTALL file to install the Perl Tk module.

APPLICATION SOFTWARE

IIR/filterTest

IIR Filter Design

The IIR filter example software presented in Chapter 7, "Digital Filters," is an application based on the user input from a perl/Tk module and a display output using Xmgrace software. Xmgrace is also being used to compute the Fast Fourier Transform from the data generated by the user interface. The data from the "filterTest.pl" are piped through the C interface "filterTest.c" to the Xmgrace display manager for the proper display of graphs.

You may compile the program "filerTest.c" using the Makefile in the directory,

- Applications/IIR/Makefile.
- Execute the program "filterTest" for the IIR filter design examples.

Applications/IMG/filterTest

IMG filtering

The IMG filter example software presented in Appendix C performs the image processing operations such as edge enhancement, edge detection, image inversion, etc. It is a perl module and requires perl/Tk interface. You can execute the software by simply entering the following command line,

- Applications/IMG/filterIMG.pl

The software reads the image file "beam.bmp" and performs the different filtering operation specified by the user input.

WAV filtering

The WAV filter example software presented in Appendix C performs the DTMF (Digital Tone Muti-Frequency) signals detection.
You can execute the software by simply entering the following command line,

- Applications/WAV/wav.pl

You may use a microphone of the PC sound card for sound input, but ideally, the tone frequencies that are sampled with an A/D converter should be passed through the filtering module.

SYSTEM REQUIREMENTS

Recommended System Requirements for the Example Software:

- 900 MHz Pentium III Processor
- Operating system installed, RedHat Linux 8.0 or 9.0.
- 2 MB of free hard disk space for the example software
- Perl version 5.8.5
- Perl/Tk module Tk-804.027
- Grace version 5.1.12
- 128 MB of Ram

- High-resolution Graphics Card
- 16x Speed CD-ROM Drive

Recommended System Requirements for Matlab Software:

- Sun Blade100
- Operating system, Solaris version 8.0
- 1MB of free hard disk space for the scripts
- Matlab Version 6.0
- High-resolution Graphics Card
- 16x Speed CD-ROM Drive

Recommended System Requirements for Scilab Software:

- 900 MHz Pentium III Processor
- Windows / 2000 / XP (Home/Pro)
- 20 MB of free hard disk space
- 128 MB of Ram
- High-resolution Graphics Card
- 16x Speed CD-ROM Drive
- Scilab version 2.7.2

Appendix E: Software Licenses

THE ARTISTIC LICENSE

Preamble

The intent of this document is to state the conditions under which a Package may be copied, such that the Copyright Holder maintains some semblance of artistic control over the development of the package, while giving the users of the package the right to use and distribute the Package in a more-or-less customary fashion, plus the right to make reasonable modifications.

Definitions:

- "Package" refers to the collection of files distributed by the Copyright Holder, and derivatives of that collection of files created through textual modification.
- "Standard Version" refers to such a Package if it has not been modified, or has been modified in accordance with the wishes of the Copyright Holder.
- "Copyright Holder" is whoever is named in the copyright or copyrights for the package.
- "You" is you, if you're thinking about copying or distributing this Package.
- "Reasonable copying fee" is whatever you can justify on the basis of media cost, duplication charges, time of people involved, and so on. (You will not be required to justify it to the Copyright Holder, but only to the computing community at large as a market that must bear the fee.)
- "Freely Available" means that no fee is charged for the item itself, though there may be fees involved in handling the item. It also means that recipients of the item may redistribute it under the same conditions they received it.

1. You may make and give away verbatim copies of the source form of the Standard Version of this Package without restriction, provided that you duplicate all of the original copyright notices and associated disclaimers.

2. You may apply bug fixes, portability fixes and other modifications derived from the Public Domain or from the Copyright Holder. A Package modified in such a way shall still be considered the Standard Version.

3. You may otherwise modify your copy of this Package in any way, provided that you insert a prominent notice in each changed file stating how and when you changed that file, and provided that you do at least ONE of the following:

a) place your modifications in the Public Domain or otherwise make them Freely Available, such as by posting said modifications to Usenet or an equivalent medium, or placing the modifications on a major archive site such as ftp.uu.net, or by allowing the Copyright Holder to include your modifications in the Standard Version of the Package.

b) use the modified Package only within your corporation or organization.

c) rename any nonstandard executables so the names do not conflict with standard executables, which must also be provided, and provide a separate manual page for each nonstandard executable that clearly documents how it differs from the Standard Version.

d) make other distribution arrangements with the Copyright Holder.

4. You may distribute the programs of this Package in object code or executable form, provided that you do at least ONE of the following:

a) distribute a Standard Version of the executables and library files, together with instructions (in the manual page or equivalent) on where to get the Standard Version.

b) accompany the distribution with the machine-readable source of the Package with your modifications.

c) accompany any non-standard executables with their corresponding Standard Version executables, giving the nonstandard executables nonstandard names, and clearly documenting the differences in manual pages (or equivalent), together with instructions on where to get the Standard Version.

d) make other distribution arrangements with the Copyright Holder.

5. You may charge a reasonable copying fee for any distribution of this Package. You may charge any fee you choose for support of this Package. You may not charge a fee for this Package itself. However, you may distribute this Package in aggregate with other (possibly commercial) programs as part of a larger (possibly commercial) software distribution provided that you do not advertise this Package as a product of your own.

6. The scripts and library files supplied as input to or produced as output from the programs of this Package do not automatically fall under the copyright of this Package, but belong to whomever generated them, and may be sold commercially, and may be aggregated with this Package.

7. C or perl subroutines supplied by you and linked into this Package shall not be considered part of this Package.

8. The name of the Copyright Holder may not be used to endorse or promote products derived from this software without specific prior written permission.

9. THIS PACKAGE IS PROVIDED "AS IS" AND WITHOUT ANY EXPRESS OR IMPLIED WARRANTIES, INCLUDING, WITHOUT LIMITATION, THE IMPLIED WARRANTIES OF MERCHANTIBILITY AND FITNESS FOR A PARTICULAR PURPOSE.

The End

GNU GENERAL PUBLIC LICENSE

GNU General Public License
Version 2, June 1991
Copyright (C) 1989, 1991 Free Software Foundation, Inc.
675 Mass Ave, Cambridge, MA 02139, USA

Everyone is permitted to copy and distribute verbatim copies of this license document, but changing it is not allowed.

Preamble

The licenses for most software are designed to take away your freedom to share and change it. By contrast, the GNU General Public License is intended to guarantee your freedom to share and change free software—to make sure the software is free for all its users. This General Public License applies to most of the Free Software Foundation's software and to any other program whose authors commit to using it. (Some other Free Software Foundation software is covered by the GNU Library General Public License instead.) You can apply it to your programs, too.

When we speak of free software, we are referring to freedom, not price. Our General Public Licenses are designed to make sure that you have the freedom to distribute copies of free software (and charge for this service if you wish), that you receive source code or can get it if you want it, that you can change the software or use pieces of it in new free programs; and that you know you can do these things.

To protect your rights, we need to make restrictions that forbid anyone to deny you these rights or to ask you to surrender the rights. These restrictions translate to certain responsibilities for you if you distribute copies of the software, or if you modify it.

For example, if you distribute copies of such a program, whether gratis or for a fee, you must give the recipients all the rights that you have. You must make sure that they, too, receive or can get the source code. And you must show them these terms so they know their rights.

We protect your rights with two steps: (1) copyright the software, and (2) offer you this license which gives you legal permission to copy, distribute and/or modify the software.

Also, for each author's protection and ours, we want to make certain that everyone understands that there is no warranty for this free software. If the software is modified by someone else and passed on, we want its recipients to know that what they have is not the original, so that any problems introduced by others will not reflect on the original authors' reputations.

Finally, any free program is threatened constantly by software patents. We wish to avoid the danger that redistributors of a free program will individually obtain patent licenses, in effect making the program proprietary. To prevent this, we have made it clear that any patent must be licensed for everyone's free use or not licensed at all.

The precise terms and conditions for copying, distribution and modification follow.

GNU GENERAL PUBLIC LICENSE
TERMS AND CONDITIONS FOR COPYING, DISTRIBUTION AND MODIFICATION

0. This License applies to any program or other work which contains a notice placed by the copyright holder saying it may be distributed under the terms of this General Public License. The "Program,"

below, refers to any such program or work, and a "work based on the Program" means either the Program or any derivative work under copyright law: that is to say, a work containing the Program or a portion of it, either verbatim or with modifications and/or translated into another language. (Hereinafter, translation is included without limitation in the term "modification.") Each licensee is addressed as "you."

Activities other than copying, distribution and modification are not covered by this License; they are outside its scope. The act of running the Program is not restricted, and the output from the Program is covered only if its contents constitute a work based on the Program (independent of having been made by running the Program). Whether that is true depends on what the Program does.

1. You may copy and distribute verbatim copies of the Program source code as you receive it, in any medium, provided that you conspicuously and appropriately publish on each copy an appropriate copyright notice and disclaimer of warranty; keep intact all the notices that refer to this License and to the absence of any warranty; and give any other recipients of the Program a copy of this License along with the Program.

You may charge a fee for the physical act of transferring a copy, and you may at your option offer warranty protection in exchange for a fee.

2. You may modify your copy or copies of the Program or any portion of it, thus forming a work based on the Program, and copy and distribute such modifications or work under the terms of Section 1 above, provided that you also meet all of these conditions:

   a) You must cause the modified files to carry prominent notices stating that you changed the files and the date of any change.

   b) You must cause any work that you distribute or publish, that in whole or in part contains or is derived from the Program or any part thereof, to be licensed as a whole at no charge to all third parties under the terms of this License.

   c) If the modified program normally reads commands interactively when run, you must cause it, when started running for such interactive use in the most ordinary way, to print or display an announcement including an appropriate copyright notice and a notice that there is no warranty (or else, saying that you provide a warranty) and that users may redistribute the program under these conditions, and telling the user how to view a copy of this License. (Exception: if the Program itself is interactive but does not normally print such an announcement, your work based on the Program is not required to print an announcement.)

These requirements apply to the modified work as a whole. If identifiable sections of that work are not derived from the Program, and can be reasonably considered independent and separate works in themselves, then this License, and its terms, do not apply to those sections when you distribute them as separate works. But when you distribute the same sections as part of a whole which is a work based on the Program, the distribution of the whole must be on the terms of this License, whose permissions for other licensees extend to the entire whole, and thus to each and every part regardless of who wrote it.

Thus, it is not the intent of this section to claim rights or contest your rights to work written entirely by you; rather, the intent is to exercise the right to control the distribution of derivative or collective works based on the Program.

In addition, mere aggregation of another work not based on the Program with the Program (or with a work based on the Program) on a volume of a storage or distribution medium does not bring the other work under the scope of this License.

3. You may copy and distribute the Program (or a work based on it, under Section 2) in object code or executable form under the terms of Sections 1 and 2 above provided that you also do one of the following:

   a) Accompany it with the complete corresponding machine-readable source code, which must be distributed under the terms of Sections 1 and 2 above on a medium customarily used for software interchange; or,

   b) Accompany it with a written offer, valid for at least three years, to give any third party, for a charge no more than your cost of physically performing source distribution, a complete machine-readable copy of the corresponding source code, to be distributed under the terms of Sections 1 and 2 above on a medium customarily used for software interchange; or,

   c) Accompany it with the information you received as to the offer to distribute corresponding source code. (This alternative is allowed only for noncommercial distribution and only if you received the program in object code or executable form with such an offer, in accord with Subsection b above.)

The source code for a work means the preferred form of the work for making modifications to it. For an executable work, complete source code means all the source code for all modules it contains, plus any associated interface definition files, plus the scripts used to control compilation and installation of the executable. However, as a special exception, the source code distributed need not include anything that is normally distributed (in either source or binary form) with the major components (compiler, kernel, and so on) of the operating system on which the executable runs, unless that component itself accompanies the executable.

If distribution of executable or object code is made by offering access to copy from a designated place, then offering equivalent access to copy the source code from the same place counts as distribution of the source code, even though third parties are not compelled to copy the source along with the object code.

4. You may not copy, modify, sublicense, or distribute the Program except as expressly provided under this License. Any attempt otherwise to copy, modify, sublicense or distribute the Program is void, and will automatically terminate your rights under this License. However, parties who have received copies, or rights, from you under this License will not have their licenses terminated so long as such parties remain in full compliance.

5. You are not required to accept this License, since you have not signed it. However, nothing else grants you permission to modify or distribute the Program or its derivative works. These actions are prohibited by law if you do not accept this License. Therefore, by modifying or distributing the Program (or any work based on the Program), you indicate your acceptance of this License to do so, and all its terms and conditions for copying, distributing or modifying the Program or works based on it.

6. Each time you redistribute the Program (or any work based on the Program), the recipient automatically receives a license from the original licensor to copy, distribute or modify the Program subject to these terms and conditions. You may not impose any further restrictions on the recipients' exercise of the rights granted herein. You are not responsible for enforcing compliance by third parties to this License.

7. If, as a consequence of a court judgment or allegation of patent infringement or for any other reason (not limited to patent issues), conditions are imposed on you (whether by court order, agreement or otherwise) that contradict the conditions of this License, they do not excuse you from the condi-

tions of this License. If you cannot distribute so as to satisfy simultaneously your obligations under this License and any other pertinent obligations, then, as a consequence, you may not distribute the Program at all. For example, if a patent license would not permit royalty-free redistribution of the Program by all those who receive copies directly or indirectly through you, then the only way you could satisfy both it and this License would be to refrain entirely from distribution of the Program.

If any portion of this section is held invalid or unenforceable under any particular circumstance, the balance of the section is intended to apply and the section as a whole is intended to apply in other circumstances.

It is not the purpose of this section to induce you to infringe any patents or other property right claims or to contest validity of any such claims; this section has the sole purpose of protecting the integrity of the free software distribution system, which is implemented by public license practices. Many people have made generous contributions to the wide range of software distributed through that system in reliance on consistent application of that system; it is up to the author/donor to decide if he or she is willing to distribute software through any other system and a licensee cannot impose that choice.

This section is intended to make thoroughly clear what is believed to be a consequence of the rest of this License.

8. If the distribution and/or use of the Program is restricted in certain countries either by patents or by copyrighted interfaces, the original copyright holder who places the Program under this License may add an explicit geographical distribution limitation excluding those countries, so that distribution is permitted only in or among countries not thus excluded. In such case, this License incorporates the limitation as if written in the body of this License.

9. The Free Software Foundation may publish revised and/or new versions of the General Public License from time to time. Such new versions will be similar in spirit to the present version, but may differ in detail to address new problems or concerns.

Each version is given a distinguishing version number. If the Program specifies a version number of this License which applies to it and "any later version," you have the option of following the terms and conditions either of that version or of any later version published by the Free Software Foundation. If the Program does not specify a version number of this License, you may choose any version ever published by the Free Software Foundation.

10. If you wish to incorporate parts of the Program into other free programs whose distribution conditions are different, write to the author to ask for permission. For software which is copyrighted by the Free Software Foundation, write to the Free Software Foundation; we sometimes make exceptions for this. Our decision will be guided by the two goals of preserving the free status of all derivatives of our free software and of promoting the sharing and reuse of software generally.

NO WARRANTY

11. BECAUSE THE PROGRAM IS LICENSED FREE OF CHARGE, THERE IS NO WARRANTY FOR THE PROGRAM, TO THE EXTENT PERMITTED BY APPLICABLE LAW. EXCEPT WHEN OTHERWISE STATED IN WRITING THE COPYRIGHT HOLDERS AND/OR OTHER PARTIES PROVIDE THE PROGRAM "AS IS" WITHOUT WARRANTY OF ANY KIND, EITHER EXPRESSED OR IMPLIED, INCLUDING, BUT NOT LIMITED TO, THE IMPLIED WARRANTIES OF MERCHANTABILITY AND FITNESS FOR A PARTICULAR PURPOSE. THE ENTIRE RISK AS TO THE QUALITY AND PERFORMANCE OF THE PROGRAM IS WITH YOU. SHOULD THE

PROGRAM PROVE DEFECTIVE, YOU ASSUME THE COST OF ALL NECESSARY SERVICING, REPAIR OR CORRECTION.

12. IN NO EVENT UNLESS REQUIRED BY APPLICABLE LAW OR AGREED TO IN WRITING WILL ANY COPYRIGHT HOLDER, OR ANY OTHER PARTY WHO MAY MODIFY AND/OR REDISTRIBUTE THE PROGRAM AS PERMITTED ABOVE, BE LIABLE TO YOU FOR DAMAGES, INCLUDING ANY GENERAL, SPECIAL, INCIDENTAL OR CONSEQUENTIAL DAMAGES ARISING OUT OF THE USE OR INABILITY TO USE THE PROGRAM (INCLUDING BUT NOT LIMITED TO LOSS OF DATA OR DATA BEING RENDERED INACCURATE OR LOSSES SUSTAINED BY YOU OR THIRD PARTIES OR A FAILURE OF THE PROGRAM TO OPERATE WITH ANY OTHER PROGRAMS), EVEN IF SUCH HOLDER OR OTHER PARTY HAS BEEN ADVISED OF THE POSSIBILITY OF SUCH DAMAGES.

END OF TERMS AND CONDITIONS

Appendix: How to Apply These Terms to Your New Programs

If you develop a new program, and you want it to be of the greatest possible use to the public, the best way to achieve this is to make it free software which everyone can redistribute and change under these terms.

To do so, attach the following notices to the program. It is safest to attach them to the start of each source file to most effectively convey the exclusion of warranty; and each file should have at least the "copyright" line and a pointer to where the full notice is found.

> one line to give the program's name and a brief idea of what it does.
> Copyright © 19yy name of author
>
> This program is free software; you can redistribute it and/or modify it under the terms of the GNU General Public License as published by the Free Software Foundation; either version 2 of the License, or (at your option) any later version.
>
> This program is distributed in the hope that it will be useful, but WITHOUT ANY WARRANTY; without even the implied warranty of MERCHANTABILITY or FITNESS FOR A PARTICULAR PURPOSE. See the GNU General Public License for more details.
>
> You should have received a copy of the GNU General Public License along with this program; if not, write to the Free Software Foundation, Inc., 675 Mass Ave, Cambridge, MA 02139, USA.

Also add information on how to contact you by electronic and paper mail.

If the program is interactive, make it output a short notice like this when it starts in an interactive mode:

> Gnomovision version 69, Copyright © 19yy name of author
> Gnomovision comes with ABSOLUTELY NO WARRANTY; for details type 'show w'.

This is free software, and you are welcome to redistribute it under certain conditions; type 'show c' for details.

The hypothetical commands 'show w' and 'show c' should show the appropriate parts of the General Public License. Of course, the commands you use may be called something other than 'show w' and 'show c'; they could even be mouse-clicks or menu items—whatever suits your program.

You should also get your employer (if you work as a programmer) or your school, if any, to sign a "copyright disclaimer" for the program, if necessary. Here is a sample; alter the names:

Yoyodyne, Inc., hereby disclaims all copyright interest in the program 'Gnomovision' (which makes passes at compilers) written by James Hacker.

signature of Ty Coon, 1 April 1989
Ty Coon, President of Vice

This General Public License does not permit incorporating your program into proprietary programs. If your program is a subroutine library, you may consider it more useful to permit linking proprietary applications with the library. If this is what you want to do, use the GNU Library General Public License instead of this License.

SCILAB LICENSE

1– Preface

The aim of this license is to lay down the conditions enabling you to use, modify and circulate the SOFTWARE. However, INRIA and ENPC remain the authors of the SOFTWARE and so retain property rights and the use of all ancillary rights.

2– Definitions

The SOFTWARE is defined as all successive versions of SCILAB software and their documentation that have been developed by INRIA and ENPC.

SCILAB DERIVED SOFTWARE is defined as all or part of the SOFTWARE that you have modified and/or translated and/or adapted.

SCILAB COMPOSITE SOFTWARE is defined as all or a part of the SOFTWARE that you have interfaced with a software, an application package or a toolbox of which you are owner or entitled beneficiary.

3– Object and conditions of the SOFTWARE license

a) INRIA and ENPC authorize you free of charge, to reproduce the SOFTWARE source and/or object code on any present and future support, without restriction, providing the following reference appears in all the copies: Scilab ©INRIA-ENPC.

b) INRIA and ENPC authorize you free of charge to correct any bugs, carry out any modifications required for the porting of the SOFTWARE and to carry out any usual functional modification or correction, providing you insert a patch file or you indicate by any other equivalent means the nature and date of the modification or the correction, on the corresponding file(s) of the SOFTWARE.

c) INRIA and ENPC authorize you free of charge to use the SOFTWARE source and/or object code, without restriction, providing the following reference appears in all the copies: Scilab (c)INRIA-ENPC.

d) INRIA and ENPC authorize you free of charge to circulate and distribute, free of charge or for a fee, the SOFTWARE source and/or object code, including the SOFTWARE modified in accordance with above-mentioned article 3 b), on any present and future support, providing:
- the following reference appears in all the copies: Scilab ©INRIA-ENPC.
- the SOFTWARE is circulated or distributed under the present license.
- patch files or files containing equivalent means indicating the nature and the date of the modification or the correction to the SOFTWARE file(s) concerned are freely circulated.

4– Object and conditions of the DERIVED SOFTWARE license

a) INRIA and ENPC authorize you free of charge to reproduce and modify and/or translate and/or adapt all or part of the source and/or the object code of the SOFTWARE, providing a patch file indicating the date and the nature of the modification and/or the translation and/or the adaptation and the name of their author in the SOFTWARE file(s) concerned is inserted. The SOFTWARE thus modified is defined as DERIVED SOFTWARE. The INRIA authorizes you free of charge to use the source and/or object code of the SOFTWARE, without restriction, providing the following reference appears in all the copies: Scilab ©INRIA-ENPC.

b) INRIA and ENPC authorize you free of charge to use the SOFTWARE source and/or object code modified according to article 4–a) above, without restriction, providing the following reference appears in all the copies: "Scilab inside ©INRIA-ENPC."

c) The INRIA and the ENPC authorize you free of charge to circulate and distribute for no charge, for noncommercial purposes the source and/or object code of DERIVED SOFTWARE on any present and future support, providing:
- the reference "Scilab inside ©INRIA-ENPC" is prominently mentioned;
- the DERIVED SOFTWARE is distributed under the present license;
- the recipients of the distribution can access the SOFTWARE code source;
- the DERIVED SOFTWARE is distributed under a name other than SCILAB.

d) Any commercial use or circulation of the DERIVED SOFTWARE shall have been previously authorized by INRIA and ENPC.

5– Object and conditions of the license concerning COMPOSITE SOFTWARE

a) INRIA and ENPC authorize you to reproduce and interface all or part of the SOFTWARE with all or part of other software, application packages or toolboxes of which you are owner or entitled beneficiary in order to obtain COMPOSITE SOFTWARE.

b) INRIA and ENPC authorize you, free of charge, to use the SOFTWARE source and/or object code included in the COMPOSITE SOFTWARE, without restriction, providing the following statement appears in all the copies: "composite software using Scilab ©INRIA-ENPC functionality."

c) INRIA and ENPC authorize you, free of charge, to circulate and distribute for no charge, for purposes other than commercial, the source and/or object code of COMPOSITE SOFTWARE on any present and future support, providing:

- the following reference is prominently mentioned: "composite software using Scilab ©INRIA-ENPC functionality";
- the SOFTWARE included in COMPOSITE SOFTWARE is distributed under the present license;
- recipients of the distribution have access to the SOFTWARE source code;
- the COMPOSITE SOFTWARE is distributed under a name other than SCILAB.

e) Any commercial use or distribution of COMPOSITE SOFTWARE shall have been previously authorized by INRIA and ENPC.

6– Limitation of the warranty

Except when mentioned otherwise in writing, the SOFTWARE is supplied as is, with no explicit or implicit warranty, including warranties of commercialization or adaptation. You assume all risks concerning the quality or the effects of the SOFTWARE and its use. If the SOFTWARE is defective, you will bear the costs of all required services, corrections or repairs.

7– Consent

When you access and use the SOFTWARE, you are presumed to be aware of and to have accepted all the rights and obligations of the present license.

8– Binding effect

This license has the binding value of a contract. You are not responsible for respect of the license by a third party.

9– Applicable law

The present license and its effects are subject to French law and the competent French courts.

Appendix F
Bibliography

"Analysis of Sallen-Key Architecture: Application Report." Texas Instruments, 2002. Found online at *http://focus.ti.com/lit/an/sloa024b/sloa024b.pdf*.

Bozic, S. M. *Digital and Kalman Filtering.* John Wiley & Sons, New York, 1979.

Department of Music, Stanford University, Stanford, CA. Found online at *http://ccrma.stanford.edu/~jos/mdft/*.

Desoer, Charles A. and Ernest S. Kuh. *Basic Circuit Theory.* McGraw Hill, 1969.

Found online at *http://www.hsdal.ufl.edu/Projects/IntroDSP/*.

Ghausi, M.S. *Electronic Devices and Circuits.* Oxford University Press, 1984.

Hamming, R.W. *Digital Filters.* Dover Publications, 1998.

Haykin, Simon. *Communication Systems.* John Wiley & Sons, New York, 1979.

Hayt, W.H., and G.W. Neudeck. *Electronic Circuit Analysis and Design.* Houghton Mifflin Company, Boston, 1989.

Karu, Zoher Z. *Signals and Systems Made Ridiculously Simple.* Zizi Press, Cambridge, MA, 1995.

Lyn, Paul A. and Wolfgang Fuerst. *Introductory Digital Signal Processing*, Second Edition. John Wiley & Sons, Chichester, 1999.

Miller, Gary M. *Modern Electronic Communication.* Prentice Hall, Englewood Cliffs, N.J., 1988.

Oppenheim, Allen V. Oppenheim and R.W. Schaffer. *Discrete-Time Signal Processing.* Prentice Hall, Englewood Cliffs, NJ, 1989.

Palacios, Jesus Olivan. "An introduction to the treatment of neurophysiological signals using Scilab." 2001. Found online at *http://www.neurotraces.com/scilab/scilab2/*.

Reid, Christopher E. and Thomas B. Passin. *Signal Processing in C.* John Wiley & Sons, New York, 1992.

Rorabaugh, C. Britton. *Digital Filter Designer's Handbook.* McGraw-Hill, 1993.

Smith, Julius O. "Mathematics of the Discrete Fourier Transform with Music and Audio Applications." Center for Computer Research in Music and Acoustics,

Smith, Steven W. *The Scientist's and Engineer's Guide to Digital Signal Processing.* California Technical Publishing, 1997.

Taylor, Fred J. "EEL 4750: Introduction to DSP." 2003.

Tolstov, Georgi P. *Fourier Series.* Dover Publications, 1962.

Who is Fourier? A Mathematical Adventure. Translated by Alan Gleason. Language Research Foundation, 1995.

Wiley Jr., C.R Wiley. *Advanced Engineering Mathematics.* McGraw-Hill.

Williams, Charles D. H. "Feedback and Temperature Control." Found online at *http://newton.ex.ac.uk/teaching/CDHW/Feedback/*.

Index

$1/f$ noise described, 97–99

A

Additive property of linear systems
   defined, 102, 103
   deriving, 105
Admittance, calculating, 157–160
Amplifiers
   inverting, 203
   noninverting, 204–205
   op-amps as, 201–203
Amplitude
   defined, 3, 4, 8
   distribution, 93–94
   finding, 20
   and periodic waves, 10–11
   sample period and, 83
Amplitude modulation (AM) described, 65
Analog filters
   converting, 263–267
   described, 184, 201–208, 223
an coefficient, computing, 22
Angular velocity defined, 9–10
ao coefficient, computing, 23, 25–26
Arbitrary waves
   defined, 11
   synthesizing, 12–16
Artistic license, 371–372
Associative property of convolution defined, 120
Autocorrelation of random sequences, 94–95, 96, 97, 99

B

Backward differences method, 111–112
Band pass filters
   characterizing, 236–239
   defined, 184, 187, 212–213
   FIR, designing, 282–284, 345
   frequency response in, 271
   realizing, 248–249
   transfer functions, 194–195, 212–213, 248
Band stop filters
   attenuation defined, 185
   characterizing, 236, 239–243
   defined, 184, 187, 218
   FIR, designing, 284–287, 345
   frequency response in, 271
   magnitude in, 239–240
   realizing, 248–249
   ripple defined, 185, 260
   transfer functions, 195–197
Bandwidth
   computing, 236–237
   controlling, 194
Bessel function, 303
bn coefficient, computing, 22
Bode plots described, 171–173, 278, 305–306, 330
Butterworth filters
   described, 197–201, 210–211, 217
   designing, 250–254, 265–267, 332, 345, 349
β value, Kaiser window, 303

C

Capacitors
   current/voltage equations, 52, 53, 55–56, 59, 103, 107
   differential equations, second-order, 123, 124
Carrier wave defined, 65
CD-ROM, about, 365–366

383

Chebyshev filters
  described, 197
  designing, 254–261, 331–332, 345–349
Chirp waveform, generating, 326
Circuits. *See also* RC Network; RLC circuits
  current/voltage relationships, 154–155
  parallel, 126–130
  Sallen-Key described, 205
  series, 127–130, 170
Coefficients
  differential equations, 243–245
  filter, generating, 281, 285, 290, 294, 297–298, 300
  in Fourier analysis
    computing, 22–23, 25–26, 73, 74
    defined, 12, 81
  frequency, fundamental, 25
  RLC circuits, 130, 138, 144
Complex numbers
  conjugates, 38–40
  in fast Fourier transforms, 88, 90
  Matlab software, 313–314
  overview, 35–36, 66–67
  in polar coordinates, 40–43
  powers of, 42
  rectangular form, converting to, 40–41
  representation of, 36–40, 47–48, 67, 141–143
  roots of, 43, 134–137, 140–141
Complex plane defined, 39–40
Continuous time
  defined, 2–3
  Fourier series described, 26–32, 70–74
Convolution process
  and arbitrary input, 116–117, 146
  described, 105, 109, 117–122, 223–224
  discrete time, 117–120, 146, 223–224
  high-pass FIR filters, 282
  and Laplace Transform, 161
  limitations of, 147
  Matlab, 328
  properties of, 120–121
  Scilab software, 342–343

Cosine (cos) functions
  amplitude of, 20
  in arbitrary waveforms, 14
  described, 6–8
  Fourier series
    continuous time, 27, 28, 33
    discrete time, 24, 33–34, 73–75
  Laplace Transformation of, 150–151
  and periodic waves, 9
  in Z-Transforms, 176
Cross correlation of random sequences, 95–96
Cumulative property of convolution defined, 120
Current
  computing, 63–64, 103
  current/voltage relationships
    capacitors, 52, 53, 55–56, 59, 103, 107
    circuits, 154–155
    inductors, 52–53, 103
  Kirchoff's Law, 106, 110, 138
  in RLC circuits, 130, 133–137
Cutoff frequency
  computing, 265
  defined, 58, 59, 185, 186, 208
  parameters, normalizing, 225

D

Damping conditions described, 124, 132, 135–137, 142–144
dc constants
  computing, 23, 25–26, 28
  in continuous time Fourier series, 27, 33
  in discrete time Fourier series, 24, 34
Decay rates in RLC networks, 123–124, 129–130
Decibels (dB) described, 170–171
Decimation filters, 278–280
Differential equations
  coefficients of, 243–245
  first-order, 106–116, 246
  form of, 105, 247–248
  and the Laplace Transform, 153–157
  linear, 104–105

overview, 101–103
second-order, 123–127
solutions defined, 104, 144
and Z-Transforms, 180–182
Differentiation described, 59
Digital filters. *See* Filters, digital
Discrete time
convolution of, 117–120, 146, 223–224
cyclic frequencies defined, 10
defined, 2–3
in Fourier series, 23, 24, 33–34, 73–75
the Fourier Transform, 85–86, 99, 329
solution, 111–112, 113
Distributive property of convolution defined, 121

E

Elliptic filters, designing, 261–262, 332–333, 345
Equations. *See specific equation by name or type*
Euler's identity
and exponent functions, 47–48, 70, 73
and nonperiodic waves, 77
Exponent functions
Euler's identity and, 47–48, 70, 73
Laplace Transformation of, 150
in Z-Transforms, 175
Exponential growth defined, 44
Exponential notation (e)
and Euler's identity, 47–48
overview, 43–51
power functions, tabulation of, 45

F

Filters
analog
converting to digital, 263–267
described, 184, 201–208, 223
characterizing, 232
converting, 235, 240–241, 263–267
designing (*See also individual filter by name or type*)
general discussion, 183–184, 186–201, 222–223, 330–331
transfer functions, 162–166, 186–187, 248, 270
digital
characterizing, 232–243
classification of, 269
converting to analog, 263–267
described, 184, 223
designing, 265–267
frequency response in, 272
transfer functions, 226–227
generating, 221–222
group delay in, 331
high-pass
characterizing, 234–235
described, 61, 187, 214–217
FIR, designing, 280–282, 345
frequency response in, 270–271
realizing, 192–194, 247
ideal, 270–274
identifying, 222
low-pass
described, 59, 162, 164, 165, 187
designing, 207–208, 232–234, 275–278, 345
frequency response in, 270–271, 272
realizing, 190–192, 209–210, 233–234, 246
order, calculating, 200
outcome, verifying, 246
output
generating, 223, 224
magnitude/phase, defining, 215–216
phase distortion, removing, 330
realizing, 232
response
analyzing, 264–265
characteristics, 186
Fourier Transform of, 223
improving, 249
realizing, 243–249
specifying, 185
terminologies, 184–186
filterTest.pl software, 246, 248, 249
Finite Impulse Response (FIR) filters

described, 269, 270
designing, 274–287, 333, 345, 350
FIR filters. *See* Finite Impulse Response (FIR) filters
Flow control, Matlab, 324–325
Forced response of systems defined, 102, 104, 108–109
Fourier analysis, 1–2, 16–17
Fourier series
    continuous time equations, 26–33, 70–74
    defined, 2, 81
    discrete time equations, 23, 24, 33–34, 73–75
    vs. Fourier Transform, 81
    frequency response, ideal, 271–273
    periodic pulse trains, 75–77
    waveforms, obtaining, 23–24, 26
Fourier synthesis, 11–16
the Fourier Transform
    computing, 77–80
    defined, 2, 11, 69–70, 77, 223
    discrete time, 85–86, 99, 329
    fast method, 86–93, 343
    of the filter impulse response, 223
    vs. Fourier series, 81
    frequency component in, 80
    and the Heisenberg Uncertainty Principle, 81, 84
    inverse, 80, 100
    rectangular window, 290
Frequency. *See also* Nyquist frequency (N)
    and angular displacement, 9–10
    complex described, 48–50
    computers, range visible to, 225
    converting, 264
    cutoff
        computing, 265
        defined, 58, 59, 185, 186, 208
        parameters, normalizing, 225
    cyclic, equations, 16, 73–74
    defined, 8–9
    doublers, designing, 66
    filters
        design of, 184, 346
        response in, 270–272
    fundamental
        coefficients, computation values for, 25
        defined, 12, 18–19
    gain and, 231–232, 235, 239
    magnitude, obtaining, 31, 227, 230
    rejection, 185
    representing, 160, 169
    response
        analyzing, 161, 166–170, 330
        described, 223
        poles/zeros and, 166–170, 172, 230
        producing, 221–222
        Transfer Function, obtaining, 228
    sampling, 31, 227, 230
    selecting, 201
    spectra, 79, 82
    and speed of execution, 173
    suppressing, 249
    and voltage calculations, 58
    waveform and, 19
    $\omega\Delta t$ variable, 224–225
    Ω sampling variable, 224–225, 226, 230, 234
Frequency-frequency mapping described, 187–189
Frequency modulation (FM) described, 65
Functions. *See also individual function by name*
    Bessel, 303
    conjugate described, 50–51
    converting, 11
    exponent, 45, 150, 175
    impulse, 150
    infinite duration, 83
    magnitude and frequency, 31, 227, 230
    orthogonal, 20–22
    periodic
        as complex number functions, 70–75
        described, 69
        representing, 271
    power, 45, 46

sinusoidal, 49–50, 52, 53
trigonometric, 21, 64–66
true *vs.* Fourier estimation, 81–84

G

Gain and frequency, 231–232, 235, 239
Gain filter defined, 185
Gaussian operator _ described, 36–37, 313
Gibb's phenomenon defined, 19
GNU General Public license, 373–378
Grace software, installing, 367
Growth function defined, 44

H

Hamming window, 300–302, 306, 333
Harmonics in Fourier analysis defined, 19, 76
Heisenberg Uncertainty Principle and the Fourier Transform, 81, 84
Heterodyning described, 66
Homogeneity of linear systems, 102, 105

I

IIR filters. *See* Infinite Impulse Response (IIR) filters
IIR-Filtertest software, 367–368
Imaginary operator _ described, 36–37, 313
Impedance, calculating, 61–62, 157–160
Impulse function, Laplace Transformation of, 150
Impulse response
    scaled, 116–117, 146
    unit
        described, 112–115, 143–145
        generating, 327
        in Z-Transforms, 175
    viewing, 223, 288–289, 292, 330
Inductors
    current/voltage relationships, 52–53, 103
    differential equations, second-order, 123, 124
Infinite Impulse Response (IIR) filters, 269, 331, 344–346
Integration described, 61

K

Kaiser window, 302–306
Kirchoff's Current Law
    and natural response, 106
    and step response, 109–111, 137–141

L

Laplace Transform
    advantages of, 153
    basic function transformations, 150–151, 154
    convolution process and, 161
    described, 105, 148–149
    and differential equations, 153–157
    differentiation, rules of, 151–152
    integration, rules of, 152–153
    inverse, 153
    linearity property of, 149–150
Lattice structures, 329
L'Hopital's Rule, 75
Licenses, software, 371–380
Linear equation systems, 320–324
Logarithmic numbers
    natural, 47
    rules of operations, 44

M

Magnitude
    in band-stop filters, 239–240
    calculating, 159, 178, 209
    Chebyshev filters, 255
    and filter output, defining, 215–216
    as frequency function, obtaining, 31, 227, 230
    power window response, 295
    rectangular window, 291
    transfer functions, 167, 168, 178, 238, 346–347
Matlab software
    arrays in, 317
    bilinear transform function, 267
    characters/text, 317

colon operator functions, 315
complex numbers, 313–314
conversions, 317–318
convolution process, 328
data, importing/exporting, 308–309
and decimation filters, 279–280
described, 307
elliptic filter function, 261–262
expressions/special constants in, 316
filter coefficients, generating, 281, 285, 290, 294, 297–298, 300
and filter modeling, 234, 250, 257
flow control, 324–325
functions, 318
linear equation systems, 320–324
matrices, operations on, 309–313, 314, 315, 321–322
polynomials, 318–320, 322–324
scripts, executing, 308–314
signal processing, 326–328
and sinc functions, 275, 276, 327
stop-band ripple function, 185, 260
structures, 318
transfer functions, 328–329
Matrices, operations on
Matlab, 309–313, 314, 315, 321–322
Scilab, 337

N

Natural response of system
defined, 102, 104, 106–108
RLC circuits, 127–137
Networks, electrical described, 51
Newton's Law defined, 29–30
Noise described, 97–99, 183
Notch filters, 240–243, 248, 249
Numbers. *See* Complex numbers; Logarithmic numbers
Nyquist frequency (N). *See also* Frequency
and cutoff frequency parameters, 225
described, 20
and discrete time Fourier series, 23, 33–34, 73

O

Op-amps described, 201–203
Order defined, 185

P

Partial fraction expansions, 329
Pass-band ripple filter defined, 185
Periodic waves
dependencies, removing, 76–77
mathematics of, 4–6
overview, 3–4
properties of, 10–11
Phase angles
changing, 235
determining, 60–61, 66
in filter design, 163, 164, 215–216
in sinusoidal functions, 49–50
transfer functions, 179, 228
Phase modulation described, 65
Phase response equation for continuous time Fourier series, 31
Phasor method
current, determining, 63–64
overview, 51–53
representation of, 54
Pink noise described, 97–98
Polar coordinates, 40–43
Poles/zeros
filter design and, 187, 331, 347–349
and frequency response, 166–170, 172, 230
transfer functions, 230–232, 233, 244, 328
of Z-Transforms, 180–182
Polynomials
Matlab software, 318–320, 322–324
Scilab software, 336–339
transfer functions, rational, 228–230, 232, 233, 235
Power functions, 45, 46
Power spectrum described, 96, 98, 99
Power window, 293–297, 298, 306
Pythagoras theorem of the right angle triangle and complex numbers, 36

Q

Q defined, 185, 186
Quality factor in RLC circuits, 142–143

R

Ramp function, generating, 327
Rational polynomial transfer functions
  described, 228–230, 232, 233, 235
RC Network
  active filters for, 205–207
  circuit schematic, 107
  filters, analyzing, 264–265
  natural response in, 106–108
Rectangular window, 290–292, 295, 296–297, 305
Resistors
  decay rates, RLC circuits, 129–130
  differential equations, second-order, 123, 124
  impedance of, 158
  voltage on, 52, 53, 60–61, 107, 158
Resonance, calculating, 62–63
Ripples, 185, 260, 286, 289
RLC circuits
  coefficients, evaluating, 130, 138, 144
  current in, 130, 133–137
  decay rates in, 123–124, 129–130
  natural response of, 127–137
  overview, 123–125
  parallel, 126–130
  quality factor in, 142–143
  series, 127–130, 170
Roll-off rate defined, 185, 209–210
Roots
  complex, 43, 134–137, 140–141
  real and distinct, 131–132, 138–139
  real and equal, 133–134

S

Sallen-Key circuit described, 205
Sawtooth waveform, generating, 326
Scaled impulse response described, 116–117, 146

Scilab software
  convolution in, 342–343
  correlation in, 343–344
  described, 335–336
  factored forms, converting, 244
  fast Fourier Transforms, 86–93, 343
  and filter modeling, 234, 250, 290
  IIR filter design, 344–346
  installing, 366
  license, 378–380
  matrices, operations on, 337
  plotting in, 339–341
  polynomial operations, 336–339
  and sinc functions, 275, 276, 327
  transfer functions in, 341
Scripts, executing, 308–314
Second order section (SOS), 329
Shift property of Z-Transforms, 176–177
Signals
  analyzing, 2
  defined, 183
  differences, amplifying, 201–203
  modulation of, 64–66
  processing, 2–3, 101–102, 105, 326–328
  random, 93–99
Sinc functions, generating, 275, 276, 327
Sine (sin) functions
  amplitude of, 20
  in arbitrary waveforms, 14
  described, 6–8
  Fourier series
    continuous time, 27, 28–29, 33
    discrete time, 24, 34
  in Fourier synthesis, 12
  Laplace Transformation of, 151
  orthogonal properties of, 21–22
  and periodic waves, 9
  in Z-Transforms, 175
Single-pole low-pass filters described, 207–208
Sinusoidal functions, 49–50, 52, 53
State space, 328–329
Step response
  generating, 327

in RC circuits defined, 109–111, 137–141
Systems
    behavior, analyzing, 173–174
    causal, 245
    control, designing, 116
    linear time invariant, 102–105, 226–227, 245
    noncausal, 245, 277
    response
        described, 29–32, 102, 104, 158
        forced, 102, 104, 108–109
        natural, 102, 104, 106–108, 127–137
        software requirements, 368–369

T
TK software, installing, 367
Transfer functions
    building blocks of, 189–190
    cascading, 245–246
    for complex conjugate pole vectors, 237
    complexities, reducing, 166
    described, 102, 156, 160–162, 170, 182
    filters
        band-pass, second-order, 194–195, 212–213, 248
        band-stop, second-order, 195–197
        design of, 162–166, 186–187
        digital, 226–227
        general discussion, 189, 273–274
        high-pass, 192–194, 247
        low-pass, 190–192, 209–210, 233–234, 246
        notch, 242
    frequency response, obtaining, 228
    magnitude, 167, 168, 178, 238, 346–347
    Matlab software, 328–329
    models of, 228
    obtaining, 30–31, 270
    phase response in, 179, 228
    poles/zero gain, 230–232, 233, 244, 328
    rational polynomial described, 228–230, 232, 233, 235
    Scilab software, 341
    in z-Transforms, 177–179
Transition band filter defined, 185

U
Unit impulse response
    described, 112–115, 143–145
    generating, 327
    in Z-Transforms, 175
Unit step functions
    with complex value s, Laplace Transformation of, 150
    Laplace Transformation of, 150
    in Z-Transforms, 175

V
Vectors
    complex, 236, 237
    magnitude and sampling frequency, 31, 227, 230
    phase angles, changing, 235
    in polar coordinates, 41
    poles/zeros as, 167
    rotating, 89, 91–92
Voltage
    in complex notation, 56
    computing, 57, 58, 103
    current/voltage relationships
        capacitors, 52, 53, 55–56, 59, 103, 107
        circuits, 154–155
        inductors, 52–53, 103
    equating, 106
    in phasor terms, 54
    on resistors, 52, 53, 60–61, 107, 158
    as time function, 109
Von Hamm window, 297–299, 306, 333

W
Waveforms
    arbitrary
        defined, 9, 11, 65
        Fourier series for, 23–24, 26
        synthesizing, 12–16
    Euler's identity and, 77
    and frequency, 19
    generating, 326–328
    periodic
        dependencies, removing, 76–77

mathematics of, 4–6
overview, 3–4
properties of, 10–11
ωΔτ frequency variable, 224–225
Ω frequency sampling variable, 224–225, 226, 230, 234
White noise described, 97–98
Windowing. *See also individual window type by name*
   described, 287–290, 333
   and FIR filters, 350
   functions, normalizing, 292
   and magnitude, 291, 295

Z

Zero input condition defined, 104
Zeros/poles
   filter design and, 187, 331, 347–349
   and frequency response, 166–170, 172, 230
   transfer functions, 230–232, 233, 244, 328
   of Z-Transforms, 180–182
Zero state condition defined, 104
Z-Transforms
   bilinear, 264
   described, 173–174, 223
   difference rule in, 176–177
   differential equations and, 180–182
   functions, basic, 175–177
   inverse, 174
   poles/zeros of, 180–182
   transfer function in, 177–179